Schrödinger's Web

Schrödinger's Web

Race to Build the Quantum Internet

Jonathan P. Dowling

Hearne Institute for Theoretical Physics
Baton Rouge, Louisiana

CRC Press is an imprint of the
Taylor & Francis Group, an **informa** business

First edition published 2020
by CRC Press
6000 Broken Sound Parkway NW, Suite 300, Boca Raton, FL 33487-2742

and by CRC Press
2 Park Square, Milton Park, Abingdon, Oxon, OX14 4RN

CRC Press is an imprint of Taylor & Francis Group, LLC

International Standard Book Number-13: 978-0-367-33761-2 (Hardback)
International Standard Book Number-13: 978-0-367-32231-1 (Paperback)
International Standard Book Number-13: 978-0-367-33762-9 (eBook)

Library of Congress Cataloging-in-Publication Data

Names: Dowling, Jonathan P., author.
Title: Schrödinger's web : race to build the quantum internet / Jonathan P. Dowling, Hearne Institute for Theoretical Physics, Department of Physics & Astronomy, Louisiana State University.
Description: Boca Raton : CRC Press, [2020] | Includes bibliographical references and index.
Identifiers: LCCN 2020016494 (print) | LCCN 2020016495 (ebook) | ISBN 9780367322311 (paperback) | ISBN 9780367337612 (hardback) | ISBN 9780367337629 (ebook)
Subjects: LCSH: Quantum computing. | Schrödinger equation. | Quantum theory.
Classification: LCC QA76.889 .D695 2020 (print) | LCC QA76.889 (ebook) | DDC 006.3/843–dc23
LC record available at https://lccn.loc.gov/2020016494
LC ebook record available at https://lccn.loc.gov/2020016495

Visit the [companion website/eResources]: [insert comp website/eResources URL]

Dedicated to My Family

Contents

Preface

This book is the sequel to my 2013 blockbuster, *Schrödinger's Killer App: Race to Build the World's First Quantum Computer.* While that book is not a prerequisite for reading this one, I will save a great deal of time by referring to long and tedious expositions in that book so that I may have room for even longer and more tedious explanations in this book. While the previous book focused on the quantum computer, this book acknowledges that quantum computers are now under rapid construction by several commercial entities such as D-Wave, Google, IBM, Rigetti, PsiQuantum, and Microsoft. The race now is to develop a quantum internet that can hook the quantum computers together and also connect many other exciting things such as quantum sensors and quantum clocks – all secured by quantum encryption. This new race is the topic of this book.

I motivate the subject for this book by discussing the launch (in August of 2016) of the Chinese quantum communications satellite Mozi. This spacecraft has demoed Space-to-Earth quantum cryptography, as well as quantum teleportation over a distance of 1,200 km. It beats the previous teleportation record of 143 km by an order of magnitude. The Japanese are also in the process of testing such quantum communications with orbiting satellites. Europe had proposed a program, unfunded, and in the U.S., we never got off the ground – literally. Canada also plans to launch a satellite soon. The U.S. and Europe are revisiting the matter in light of Chinese success.

As with the classical internet, a long-haul quantum internet will consist of nodes connected by communications links that span the globe, both on earth and in space, and when you talk long-haul communications, you're talking photons. We'll begin by discussing the nature of light. We'll move into a recap of how distant particles of entangled light – Schrödinger's Web – are generated and what are their uses, particularly for quantum imaging, metrology, and sensing. Then we'll discuss teleportation and quantum cryptography, then distributed quantum computing, and finally wind up with ideas about a global network of satellites and ground stations all linked together – the quantum internet.

If you are talking about either the classical or quantum internet, then you're talking about transmitting data over long distances using light in optical fibers. To understand the traditional web, you will have to follow the classical theory of light. In chapter one, "Many Hands Make Light Work," we review the development of the classical theory from the mid-1600s up until the late 1800s. Now, if you're talking about a quantum internet, then you are not talking about light waves, but instead, you're talking about photons. Photons are quantum mechanical beasts with no classical analog. Thus, in chapter two, we develop the quantum theory of light – with all of the Schrödinger-cat like unreality, uncertainty, and nonlocality. Photons are the particles that will power the quantum internet, so we had better be sure we know what the heck they are. I devote the entirety of chapter three to the notion of entangled photons. Once that is in hand, in chapter four, I discuss the unique components of a quantum network, and in chapter five, how to put them together. Penultimately, I will go utterly off-script in chapter six and layout applications for the quantum Internet for the next 100 years or so. Finally, in chapter seven, I will discuss networks of quantum sensors – which is really a hot new topic.

Foreword

[1]

1 Permission to use this drawing from Zach Wienersmith @ZachWeiner. Please format so the comic takes up 2–3 pages like a short comic book.

"When I read Dowling's previous book, *Schrödinger's Killer App*, I was inspired to draw this cartoon. I can't wait to see what I will draw after reading his new book!" – Zach Wienersmith, creator, *Saturday Morning Breakfast Cereal Comic.*

Author

Jonathan Dowling earned a Ph.D. in Mathematical Physics from the University of Colorado. He has worked at the United States Army Aviation and Missile Command, the NASA Jet Propulsion Laboratory, and now at the Louisiana State University. Dr. Dowling is one of the founders of the U.S. Government program in quantum computing and quantum cryptography.

CHAPTER 1

Many Hands Make Light Work

[1]

In the beginning, God created the heaven and the earth. And the earth was without form and void, and darkness was upon the face

1 Sistine Chapel ceiling, "The Separation of Light and Darkness," by Michelangelo di Lodovico Buonarroti Simoni. Work is in the public domain.

> of the deep. And the Spirit of God moved upon the face of the waters. And God said, "Let there be light," and there was light. And God saw the light, and it was good, and God divided the light from the darkness.[2]

THE DARKENED ROOM

The current internet (and the future quantum internet) runs on light. What is light? This is a question that humans have been asking themselves for thousands of years, with the first written record in western culture coming from *The Book of Genesis*. A precursor to the modern camera (the pinhole camera) may have been discovered accidentally as far back as 30,000 years ago. The pinhole camera is a darkened box with a pinhole poked in it. I made one of these out of a cylindrical, cardboard, Quaker Oats container as an undergrad in a beginning physics lab. I placed a circular piece of undeveloped photographic film at one end of the tube (in a dark room), taped a piece of aluminum foil over the other end, and then poked a hole in the tinfoil with a pin. I took the contraption up to the roof of the physics building, pointed the pinhole at the University of Texas tower, opened the cardboard shutter, counted to ten, shut it up and went back down, and developed the film. There appeared a ghostly image of that infamous owl-faced tower.[3]

There are theories that ancient cave drawings were inspired by such images, shining through small cracks in the cave wall and into the cave itself, where the image would appear (upside down) on the opposite wall. All the inhabitants had to do is to trace the image with coal or other pigments. Once they developed artistic skills of their own, the theory goes, they had no further need for the pinholes.

Such pinhole cameras, making images of the sun, have been described in ancient Chinese writings as far back as 1046 BCE. Pinhole cameras are one of the safest ways to view a solar eclipse or a sunspot – as the image of the eclipse is projected onto a sheet of paper and not on the back of your eyeball where it could fry your retina. (Never look directly into the sun.[4])

The first details and accurate written descriptions of the pinhole camera and the science behind the imaging process go back to the writings of the

2 Genesis 1.5, King James version of the English Bible. Work is in the public domain.

3 The real story of the owl in the University of Texas clock tower, https://alcalde.texasexes.org/2012/01/legendary-ut/.

4 "Yes, Donald Trump really did look into the sky during the solar eclipse," by Chris Cillizza, CNN (2017), www.cnn.com/2017/08/21/politics/trump-solar-eclipse/index.html.

Chinese Philosopher, Mozi, from around 500 BCE. He understood that the pinhole acted like a lens, and his drawings of the camera are similar to what you would find in a modern book of optics. The name Mozi became synonymous with science and the study of light in Chinese culture. As a foreshadowing of where I am going with this, there is now (as I type) a Chinese quantum optical communications satellite named Mozi. It is currently orbiting the earth, and its handlers just announced today (Friday 16 June 2017), the results of experiments transmitting quantum entangled states of light from the satellite to the ground. The spacecraft uses a transmitter that produces a rainbow of different-colored photons – what I call "Schrödinger's Rainbow" (Figure 1.1).

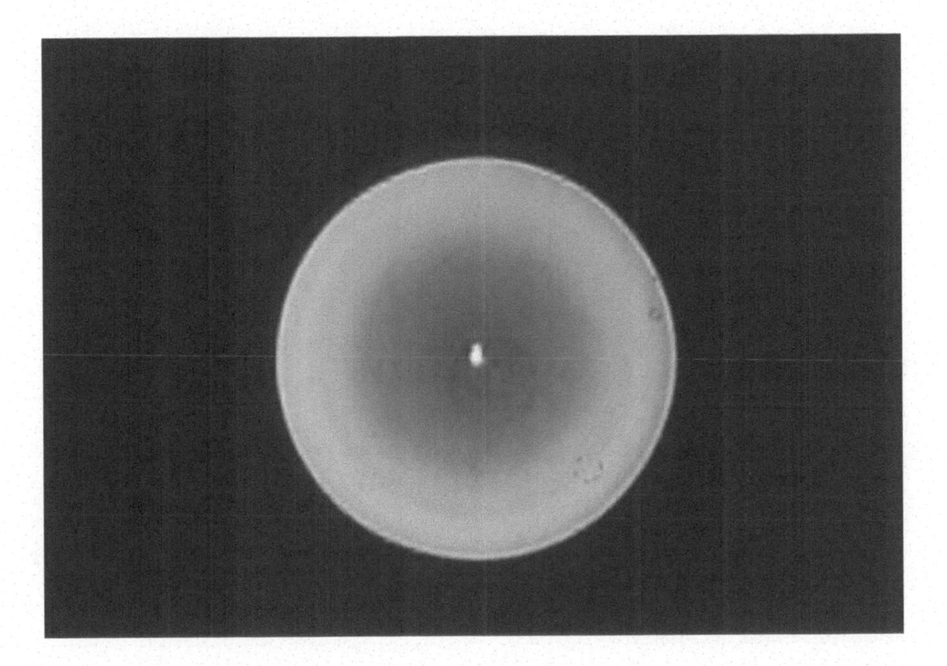

FIGURE 1.1 Schrödinger's Rainbow – separating the light from the darkness. Photo of quantum entangled photons produced in my former group's laboratory at the NASA Jet Propulsion Laboratory, circa 2003.[5]

5 Photo of entangled photons generated by spontaneous parametric down-conversion, taken by Dimitry Strekalov, circa 2003, in our lab at the NASA Jet Propulsion Laboratory. The photo was taken by a government contractor in the course of this work duties and not copyrightable under U.S. law. For details on how this image was created see, https://en.wikipedia.org/wiki/Sponta neous_parametric_down-conversion.

An older name for these pinhole cameras, used for popular entertainment in Europe hundreds of years ago (some can still be found today[6]), was a *camera obscura*, which is Latin for a darkened room. These devices were human-sized darkened rooms with a hole punched in the side. You would locate actors in daylight outside the chamber, and the gap would project ghostly images of them inside onto the wall (upside down) for the entertainment of the audience in the room. Mozi understood that the image became inverted as a result of ray tracing the light beams (Figure 1.2). In the 1800s, when we invented photographic film, we shrunk the camera obscurae (darkened rooms) down to the size of a shoebox. The images became permanently written into photographic film, which, when developed, gave you a permanent shot of the object or person (just like with my oatmeal box).

The shoebox devices were still called *camera obscurae*, but eventually, they became merely known as cameras (because nobody could pronounce *obscurae*). Hence our English word "camera" simply means "room". One problem with the pinhole camera is that the pinhole lets very little light through, and so the image takes a long time to record on the photographic film. Small children, having their portraits taken, would have to sit completely still for minutes. Photographers developed elaborate clamps to attach to the sides of their heads under their hair to keep them from moving. Hence this camera was not ideal for use on toddlers in their terrible twos (and the clamps possibly amounted to child abuse). Camera enthusiasts started making the pinholes bigger to let in more light, to shorten the exposure time, but the image was out of focus. So, they added glass lenses to compensate and to make the image arrive at the viewfinder and the film right-side up. In the 1990s and 2000s, we finally replaced the film with charge-coupled devices, and the modern digital camera was born.[7] Interestingly, on most cellphone cameras, they still use pinholes and not lenses because the digital detecting arrays are very sensitive to even feeble rays of light, and there is not much room for lenses in your ultra-slim iPhone.

If we look at how we trace the rays of light in Figure 1.2, we see how the beams cross over each other at the pinhole, and so it would appear that light moves in perfectly straight lines – like machine-gun bullets. It

6 You can visit one of the few remaining large camera obscurae in Bristol, UK. See https://en.wikipedia.org/wiki/Clifton_Observatory.

7 Charge-coupled device https://en.wikipedia.org/wiki/Charge-coupled_device.

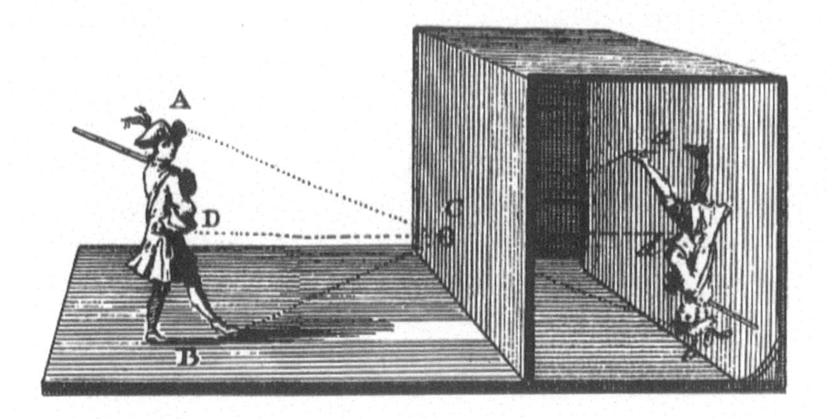

FIGURE 1.2 Schematic of an 18th-century camera obscura. The dotted lines trace three beams of light, showing that the soldier's hat (A) is mapped to the bottom of the camera, his shoes (B) to the top, and his belt (D) to the center (C), demonstrating that the image becomes inverted. The darkened room is to the right (where the audience sat), and the pinhole is located at C.[8]

would be very natural to assume that is what light consists of - particles akin to swiftly moving bullets. That view agrees with this diagram and even agrees if you start adding lenses and mirrors (at least up to a point).[9] From such observations, a popular point of view evolved that light was composed of fleet little bullets of energy that moved in very straight lines. This bullet idea was called the particle theory of light, and a famous English scientist, apple-fell-on-his-head Isaac Newton, was its's champion.[10]

NEWTON'S BULLETS

The challenge facing Newton was to get all his optics experiments, carried out in the late 1600s, to agree with his particle theory. Newton had developed an entire approach to mechanics where objects such as apples and the earth could be treated as point particles, so why not light? To explain how light

8 Figure from https://en.wikipedia.org/wiki/Camera_obscura#/media/File:001_a01_camera_obscura_abrazolas.jpg. Author unknown. This work is in the public domain in its country of origin and other countries and areas where the copyright term is the author's life plus 70 years or less. This work is in the public domain in the United States because it was published (or registered with the U.S. Copyright Office) before 1 January 1923.

9 For more information on the ray tracing of light, see https://en.wikipedia.org/wiki/Ray_tracing_(physics).

10 You can find a biography of Newton here https://en.wikipedia.org/wiki/Isaac_Newton.

moves through optical devices with lenses such as microscopes, telescopes, and eyeglasses, a simple ray-tracing model of the light particles suffices. But Newton had also done experiments with prisms and other optical elements to demonstrate that white light is composed of many colors. Newton was the first to write down a mostly correct scientific theory of the rainbow. His idea was that the varying hues of light corresponded to different particles of light moving at different speeds in the water droplets in the sky, causing the rainbow. He was able to predict the position and the angular opening of and color dispersion of the primary rainbow, but not that of the less-often-seen secondary rainbow where the colors are reversed.[11] His theory also could not explain Alexander's dark band (not a heavy metal group) between the inner and outer rainbow. (See Figure 1.3.)

FIGURE 1.3 Primary and secondary rainbows with Alexander's dark band between them. Newton's theory could not explain the dark band or why the colors in the outer rainbow are in reverse order.[12]

11 A scientific history of the development of the theory of the rainbow can be found here https://en.wikipedia.org/wiki/Rainbow#Scientific_history.

12 Figure from https://en.wikipedia.org/wiki/Rainbow#/media/File:Regenbogen_%C3%BCber_dem_Lipno-Stausee.JPG, "Alexander's band visible between the primary and secondary bows," by Alexis Dworsky. This file is licensed under the Creative Commons Attribution 2.0 Germany license.

To decode those features required an understanding of wave interference, which would not come along for about 100 years with the work of Fresnel. Even worse, Newton's theory could not convincingly explain the polarization of light. Until he died, he was plagued by his experiments with Icelandic spar. He published all of his work in 1706 in his book *Opticks*, which became the bible of the particle theory of light.

It became clear in Newton's time that light was polarized, and that a beam of light of a single color could be broken into two more beams based on its polarization. This effect is the same that you get when you look at the surface of a lake with polarizing sunglasses. The glasses cut out about half the sunlight that is giving rise to glare and allows you to see deep into the water for fish.[13] Such eyewear also cuts down on the glare reflecting off oncoming cars, and we use specialized polarizing lenses to give the illusion of three-dimensional images in 3D films. All these devices work on the principle that monochromatic light comes in two types that can be separated by a polarizing device. What is this extra property of light that gives rise to this effect? If light is a particle, as Newton insisted, then each particle had to have some additional degree of freedom that allowed this separation. Newton hypothesized that the light particles came in two types – that they had poles like the north and south poles of a magnet. In this way, perhaps, you could separate the north poles from the south poles by using suitable polarizing instruments. Alas, the particle theory could not wholly explain how light moved through the dreaded Icelandic spar.

Icelandic spar, more commonly known as a calcite crystal, is used to demonstrate the polarization of light. As mentioned above, the light that reflects off a car hood or a lake is polarized. Sunlight coming from the sky is also polarized because the rays are scattered off the molecules in the atmosphere. This scattered light is blue, which is why the sky looks blue even if you are not directly looking toward the sun. The light is maximally polarized at a point in the heavens at a right angle to the sun. At sunset or sunrise, the blue sky will be maximally polarized directly overhead.

We can see this polarization with the naked eye[14] and a polarizing material – and Icelandic spar is such a polarizing material. According to

13 For more on polarizing eyeglasses, see https://en.wikipedia.org/wiki/Polaroid_Eyewear.

14 The human eye can see polarized light via a biological effect called Haidinger's brush. As the story goes, the spar acted as the first polarizer, and the eye served as the second. By rotating the spar and staring at one point on the crystal, that point would appear to get light and dark. Try it by replacing the spar with a polarizing lens (from an old pair of sunglasses) and rotating that while looking at the sky. For more on Haidinger's brush, see https://en.wikipedia.org/wiki/Haidinger%27s_brush#Physiological_causes.

Ol' Ike Newton, sunlight that reflects off the lake is polarized one way, and the sunlight that is transmitted to the bottom is polarized the other way – typically, we now say one is horizontally polarized and the other vertically. However, light in the Icelandic spar takes *two* paths within the crystal, with one track polarized horizontally (the North Pole) and the other vertically (the South Pole). Newton insisted, without proof, that the light particles had these poles, and the different poles

FIGURE 1.4 The lines from the graph paper appear to double as the light passes through the calcite crystal. This phenomenon is called birefringence, which is seen in Icelandic spar but not in quartz or glass. One half of the doubled lines are horizontally polarized, and the other half are vertically polarized. If you put another polarizer on top of this one and rotate it, the lines will blink in and out.[15]

15 Graphic from https://en.wikipedia.org/wiki/Birefringence#/media/File:Crystal_on_graph_paper.jpg, "A calcite crystal laid upon a graph paper with blue lines showing the double refraction," by APNMJM. This file is licensed under the Creative Commons Attribution-Share Alike 3.0 Unported license.

caused this separation in the two trajectories. But why did the two different polarizations take different paths in only Icelandic spar and not in anything else like quartz or glass? No one knew. The theory was not all that convincing – but Newton was! – and for about 100 years, Newton's was the dominant explanation. Icelandic spar be damned!

As the name suggests, Icelandic spar is found in Iceland and has a fascinating history going back to the time of the Vikings. Writings of those horned-hat people show that they used Icelandic spar (which they called sunstone) to navigate their ships in the North Atlantic, even when the sky was so overcast it was impossible to see where the sun was (hence the name sunstone).

The calcite crystal has a whacky property that, if you put it on top of a sheet of paper with writing or lines on it, the script or lines will split in two – as if you were seeing double (Figure 1.4). If you put a second chunk of calcite on top of the first (or the lens from a polarizing pair of sunglasses) and rotate it around, even more bizarrely, one image would disappear. In contrast, the other remained, and if you kept turning the top crystal, the second image would go, and the first would return. Somehow the two beams of light coming from the picture were separated (seeing double) and polarized (each having a separate pole or horizontal–vertical property).[16] This strange phenomenon of light beams splitting in two in a crystal is called birefringence. The word means that the light of two different polarizations refracts into two separated beams.[17]

The Vikings discovered that by rotating the spar around and around, the sky behind it would appear to get bright or dark, even if they could not see the sun at all. Using this phenomenon, the Vikings could tell where the sun was on a cold and misty and overcast Icelandic day. Knowing the location of the sun is critical for navigation in overcast

16 For a video on the weird optical properties of calcite (Icelandic spar), see www.youtube.com/watch?v=t_w0OmgJzd0.

17 For the wave theory of birefringence in calcite, see https://en.wikipedia.org/wiki/Birefringence.

FIGURE 1.5 A disembodied hand uses a piece of Icelandic spar to determine the location of the sun.[18] As the hand rotates the spar, the clouds turn dark and light as different polarizations are filtered in and out. When the sky looks darkest, you know the sun is either on the horizon to your left or your right. If you have some idea of the time, since the sun is only on the horizon at sunset or sunrise, this gives you enough information to find the sun.

places, and this trick allowed the Vikings to stumble onto North America.[19]

What was clear is that if you wanted a complete theory of light, you had to explain the puzzling behavior of it as it passed through the Icelandic spar. Newton's theory of birefringence was not entirely satisfactory, and

---

18 Graphic from https://en.wikipedia.org/wiki/Iceland_spar#/media/File:Silfurberg.jpg, "Iceland spar, possibly the Icelandic medieval sunstone used to locate the sun in the sky when obstructed from view," by ArniEin. This file is licensed under the Creative Commons Attribution-Share Alike 3.0 Unported license.

19 A history of the Icelandic spar is here https://en.wikipedia.org/wiki/Iceland_spar; A story about it being used by Vikings to navigate their ships is here www.sciencemag.org/news/2011/11/viking-sunstone-revealed.

many other scientists raised objections to it immediately. The foremost objector was Newton's nemesis – Christiaan Huygens. (Never ask a native speaker of the Dutch language to pronounce the name Christiaan Huygens without first wrapping your head in a towel.)[20]

WAVES OF HUYGENS' – PTOOEY!

Christiaan Huygens[21] was a Dutch scientist who had a radically different theory of light that he called wavelets. His theory was the competing theory to Newton's particles. Huygens imagined that light consisted of little waves that moved in spherical patterns off of each point in space like the ripples on a pond.[22] While a bit more challenging to deal with mathematically, it too was able to explain polarization and the separation of the rainbow into colors and the opening angle of the smaller primary rainbow, but it also had problems with the sizeable secondary bow with its reversed colors, the intervening Alexander's dark band, and the bane of Icelandic spar. In spite of calling them wavelets, in Huygens *Treatise on Light* (published in 1690 – some years before Newton), it was clear he did not yet understand the concept of interference that would explain these other effects.[23] (See Figure 1.6.) Since Huygens was not nearly as famous as Newton, the scientific community mostly ignored his theory. In the end, it was about as successful (or unsuccessful) as Newton's theory, but it required a lot more work to use. However, these two different theories left open a big question in everyone's mind – is light more like a particle or a wave? There was no

20 Back in the early 2000s, I was having a whiteboard discussion at NASA JPL with my postdoc Pieter Kok (now a professor at the University of Sheffield in the UK). I began to draw a figure of Huygens' wavelets and said, "Let's consider Huygens' principle." (I pronounced the name "Hoy-Gans.") Pieter exclaimed, "*Who!?*" I replied, you know, the famous Dutch scientist Huygens and I wrote the name on the board. Pieter retorted, "You mean [insert sound of a cat hacking up a hairball with plenty of spit]!?" Years later, when I wrote him a letter of reference for his current job, I put in the letter, "Pieter's English is excellent – probably better than mine – but never ask him to pronounce the name of the famous Dutch scientist Christiaan Huygens!" At his interview, one of the interviewers took the bait, went to the board, wrote down the offending name, and said, "One last question – how do *you* pronounce *this* name?" Pieter knew I had set him up. In spite of the interviewer not wearing a towel, Pieter got the job anyway.

21 A biography of the unpronounceable Dutch scientist is here https://en.wikipedia.org/wiki/Christiaan_Huygens.

22 More on the Huygens–Fresnel principle and the history, https://en.wikipedia.org/wiki/Huygens%E2%80%93Fresnel_principle.

23 An English and spit-free version of Huygens' Treatise on Light is here https://archive.org/details/treatiseonlight031310mbp; The history leading up to this work is here https://en.wikipedia.org/wiki/Treatise_on_Light.

experiment in 1700 that could tell between these two theories. It took another hundred years after Newton and Huygens to figure this out and another hundred years to unfigure it out again.[24]

OUT – DAMNED SPOT!

The problem in physics is that you can't distinguish between two theories – such as Newton's particle theory and Huygens' wavelet

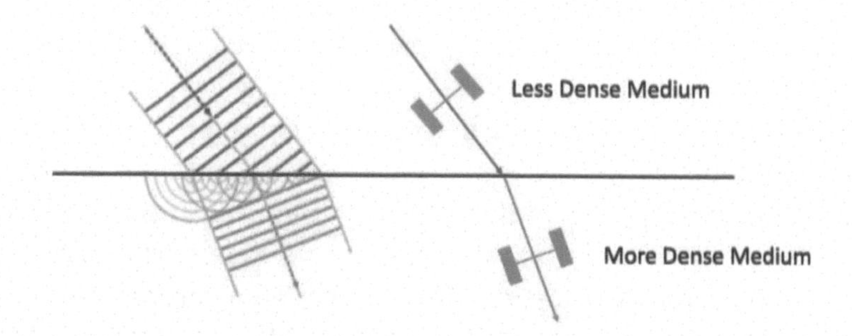

FIGURE 1.6 Huygens' principle at work. Left: A beam of light entering the water (below) from the air (above) bends downward in a process known as refraction.[25] Refraction occurs because light rays move slower in water than in air and the whole beam twists downwards away from the water's surface. Right: This effect is similar to that of driving your car at an angle off the road (less dense) into the sand (denser). The front right wheel suddenly moves much slower in the sand while the front left wheel continues at a higher speed causing the car to turn rightward. To explain this effect, Huygens treats each point at the surface of the water (dots on the left) as spherical emitters. These wavelets all add up to give the beam of light a new direction. The particle explanation of Newton looks a lot more like the car on the right where the different wheels are different particles. This process of refraction explains why when you throw a spear at a fish in a pond, you should aim not where you see the fish, but slightly below where you look at the fish.[26]

---

24 Spoiler alert! It turns out, according to quantum theory, light is both a wave and a particle at the same time – like Schrödinger's cat being dead and alive at the same time!

25 For more on the theory of optical refraction, including tips on fishing, see https://en.wikipedia.org/wiki/Refraction.

26 The figure is a modification of this https://en.wikipedia.org/wiki/Huygens%E2%80%93Fresnel_principle#/media/File:Refraction_-_Huygens-Fresnel_principle.svg, "Wave refraction in the manner of Huygens," by Arne Nordmann. This file is licensed under the Creative Commons Attribution-Share Alike 2.5 Generic, 2.0 Generic and 1.0 Generic license, which allows me to modify the work.

theory – unless you perform an experiment in the lab that rules one in and the other out. The problem with these two fundamental theories, circa 1700, was that they both seemed to agree well on the same experiments and disagreed just as poorly in other tests. There was no real difference in the observational predictions, and so it was a matter of taste as to which theory you preferred. However, things began to get heated in the early 1800s when scientists began to improve on the wavelet theory to the point where it made predictions that disagreed with the particle theory. They were no longer the same theory, and one could go to the lab and test which one was right. As usual, the experimentalists were to blame for starting the Battle of Poisson's Spot, which was not a spot on a speckled trout.

Around 1800, an idea that light did not travel in straight lines, as the Newton and Huygens theories suggested, began to take hold. A critical experiment was carried out in 1807 by an English – *gasp!* – medical doctor and part-time scientist named Thomas Young.[27] His experiment seemed to disprove the particle theory of Newton – the Pope of English science – heresy! Young had written papers on the theory of sound waves and water waves passing through holes, slits, and around objects, as well as carrying out experiments based on this theory. He made the bold assertion that light propagated much more like a water wave than a particle, as Huygens had proposed. But unlike Huygens, Young was the first to insist that light waves, like water waves, should interfere – where peaks meeting peaks made a more prominent peak – but peaks meeting troughs made nothing at all. This effect, observed in water waves moving across a tranquil pond, is called constructive and destructive interference, respectively. It is the calling card of true wave propagation.[28] Huygens had entirely missed this interference possibility with his wavelet theory. Armed with this analogy to water and sound waves, Young carried out a groundbreaking experiment with light, which is now called Young's two-slit (or double-slit) diffraction experiment.[29]

In Figure 1.7, we see a sketch of the experiment and some actual photos of the bright and dark bands of light that were predicted by

---

27 Young's biography is here www.britannica.com/biography/Thomas-Young.

28 For more on Young and Fresnel's theory of wave interference, see https://en.wikipedia.org/wiki/Interference_(wave_propagation).

29 History of Young's two-slit interference experiment with light, https://en.wikipedia.org/wiki/Young%27s_interference_experiment.

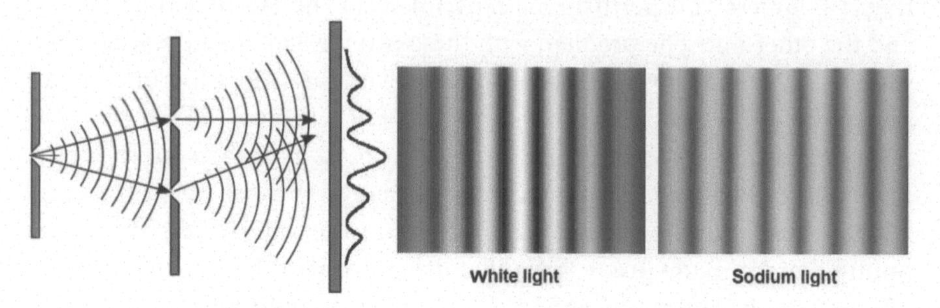

FIGURE 1.7 On the left is a sketch of the two-slit experiment inspired by water waves. Water passes through the left slit, then the waves move through the two slits in the middle, and finally, the interfering troughs and peaks can be seen in the figure on the left. Young argued the same setup would work for light – if light were a wave. Instead of troughs and peaks, you would see dark bands (gutters) and bright bands (peaks) of light on the photographic screen. The photos in the figure on the right show these bands for white sunlight, as Young used, and also for single-colored yellow light from a sodium lamp.[30]

Young. Since water moved like a wave, and light showed interference like water, the conclusion was that light was also a wave. You might think scientists would congratulate Young for such a fascinating experiment. But remember this upstart Englishman just proved that the Lion of Physics, Isaac Newton, had no claws, at least when it came to his particle theory of light. If Young was correct, Newton had to be wrong. The notion that Sir Isaac was wrong did not sit well with Newton's proponents of the particle theory, and they desperately tried to explain Young's results with particles (while simultaneously claiming Young's results were just wrong).[31] To see how wrong Newton's theory had to be, take a gander at Figure 1.8. If light traveled like bullets out of

30 Figure adapted from https://commons.wikimedia.org/wiki/File:Young%27s_two-slit_experiment_and_Lloyd%27s_mirror.png, "A drawing of Young's two-slit experiment is on the left. Diffraction patterns from white light and sodium D light are [right] …," by Stannered, Epzcaw, and Stigmatella Aurantiaca. This file is licensed under the Creative Commons Attribution-Share Alike 3.0 Unported license.

31 "Everything you have presented here is completely wrong, and besides, it's all in my book!" – Johann Rafelski. "Is that so? Well, what else in your book is completely wrong!?" – Jonathan P. Dowling. (During the questions at the end of Dowling's talk on self-field quantum electrodynamics at a physics conference in Maratea, Italy, in June of 1987. Rafelski was at the time a distinguished professor and Dowling was at the time an extinguished PhD student.)

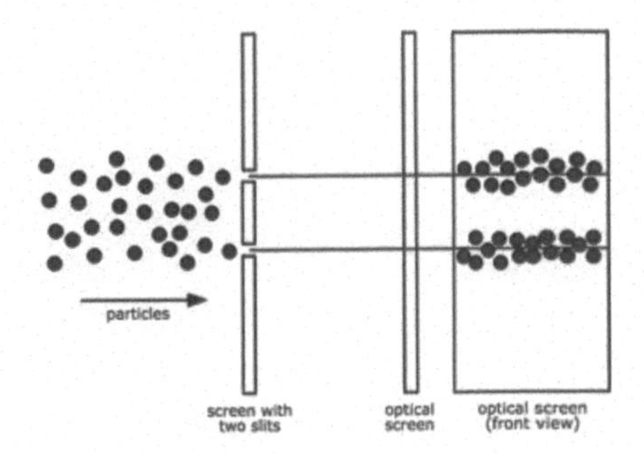

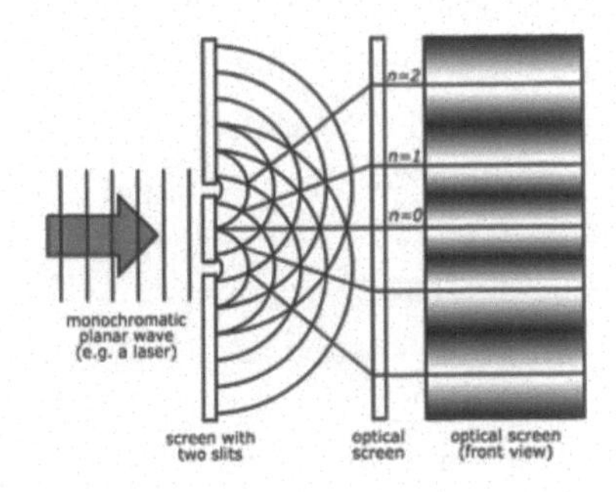

FIGURE 1.8 In Newton's bullet theory of light, on the left, instead of the alternating bright and dark bands seen in the wave experiment, you would get two bright bands right after the slits. This effect is not what Young observed. Instead, Young's wave theory, on the right, predicts that you will see many periodically alternating bands of light and dark. That is what Young saw in his experiment. Hence Newton's theory had to be wrong.[32]

a machine gun, instead of interference (many dark and light bands), you would get two bright bands behind the slits, as the light bullets moved in straight lines without interfering, shown on the left. You would never get the interference bands of waves, shown on the right.

You might think, at least after publishing his wave theory and experiment in the prestigious *Proceedings of the Royal Society of London*, scientists would hail Young as the hero of optics. Instead, he was cursed at and reviled as the villain of optics. It was as if Young were a Catholic Priest who walked into the Vatican at High Mass – and declared that the Pope was *wrong*! (In this case, the Pope was Newton, and the Vatican was the Royal Society of London.)[33] To be fair, Young's theory was more of an argument by analogy than a logical approach, and there is some doubt if he even experimented! This gap left him open to attack from the large number of particle proponents left in

32 This is a compilation of two graphics taken from https://commons.wikimedia.org/wiki/File:Experience_des_deux_fentes_(exp%C3%A9rience_des_trous_d%27Young)_avec_de_la_lumi%C3%A8re.svg and https://commons.wikimedia.org/wiki/File:Exp%C3%A9rience_des_trous_d%27Young_avec_des_particules.svg. Both files are licensed under the Creative Commons Attribution-Share Alike 3.0 Unported license.

33 "On the theory of light and Colours," by Thomas Young, M. D. F. R. S., Phil. Trans. R. Soc. Lond. **92**, 12–48, published 1 January 1802. An online version is here http://rstl.royalsocietypublishing.org/content/92/12.full.pdf+html. (I, *myself*, have a paper published in *this* journal!)

Newton's vast wake.[34] (Newton died in 1727 nearly 30 years before Young's paper came out.) Then along came French theoretical physicist, Augustin-Jean Fresnel, to further stir the wave-theory pot.[35]

Motivated by Young's experiment, Fresnel took up where Young and Huygens had left off. As we recall, the original Huygens theory still predicted that light moved in straight lines and was, for all practical purposes, equivalent to Newton's approach in its predictions. While Huygens' wavelet theory could explain reflection and refraction, it could not explain diffraction – the spreading of light into bright and dark bands after it passed through a slit. Young was the first to see the missing ingredient, that of wave interference, but his arguments were not rigorous. Fresnel cleaned this all up and put the wavelets – with diffraction – on a firm mathematical basis in what is now known as the classical wave theory of light. We call this approach the Huygens–Fresnel theory, but I think the Huygens–Young–Fresnel method would be more generous to Young. Fresnel's prize-winning book, *Memoir on the Diffraction of Light*, appeared in 1819, over a hundred years after Huygens' treatise (but only a few years after Young's paper).[36] It was a full-frontal assault on the particle theory and, because it was a rigorous mathematical treatment, there was enough red meat on it for the particle sharks to start a feeding frenzy. The Fresnel theory not only confirmed Young's experiment but could be used to predict entirely new tests that the particle theory had no chance to compete with. The first particle shark to go in for the kill was the French mathematician Siméon Denis *Poisson* (rhymes with something fishy going on).[37]

Poisson made many scientific contributions to probability theory, the theory of gravitational fields, electric fields, and Newtonian mechanics. Perhaps because of the last of these contributions, he was the most vociferous defender of Newton's particle theory of light. Poisson was the

---

34 *The Last Man Who Knew Everything – Thomas Young, the Anonymous Genius Who Proved Newton Wrong and Deciphered the Rosetta Stone, Among Other Surprising Feats*, by Andrew Robinson (Oxford University Press, 2007) pp. 123–124, which can be found via www.worldcat.org/oclc/973636581.

35 Fresnel's biography, https://en.wikipedia.org/wiki/Augustin-Jean_Fresnel.

36 "Mémoire sur la diffraction de la lumière," by A. Fresnel, (deposited 1818, "crowned" 1819). In *Oeuvres complètes*, vol.1, pp. 247–363. Paris: Imprimerie impériale, 1866–70 ; partly translated as "Fresnel's prize memoir on the diffraction of light." In *The Wave Theory of Light: Memoirs by Huygens, Young and Fresnel*, edited by H. Crew, pp. 81–144, American Book Co., 1900, archive.org/details/wavetheoryofligh00crewrich.

37 More on Mssr. Poisson, https://en.wikipedia.org/wiki/Sim%C3%A9on_Denis_Poisson; "Poisson" is the French word for "fish."

head of the particle-shark feeding frenzy at the Royal Academy of Sciences of France (RASF). He and fellow shark schoolmates were aghast at the wave theory and the experiment of Thomas Young. They were even more appalled that Young's work had started to convert previously particle-positing shark pups into tentative adoptees of the wave theory.[38] Poisson decided to put an end to this nonsense once and for all – but instead, he ended up biting off his own tail. In 1818, some ten years after Young's publication, Poisson and his buddies goaded the RASF into announcing a competition – with a substantial cash prize – for the best paper that could explain the known properties of light. Poisson fully expected the winner to be a particle proponent, and he saw his chance to kill off the wave theory once and for all. He knew Fresnel would submit his paper on the wave theory to the competition. Since Poisson was on the judging panel for the prize, he ripped into Fresnel's article with a bloody relish and decided to take it down with a strawman argument – the last refuge of an illogical scoundrel.

Fresnel's theory purported to be a general theory of the diffraction and the subsequent interference of light waves either through a hole or around an opaque object. Chewing on the latter idea, Poisson sought to wield Fresnel's theory against him. Poisson set up the following design for a thought experiment. Instead of striking two slits, as in Young's investigation, Poisson made the light hit a black circular disk. According to the particle theory, the disk would cast a shadow on a screen that would be entirely dark in the zone behind the disk – as when the moon eclipses the sun. However, the Fresnel theory predicted that, when the wavelength of the light (500 nanometers for red light) was somewhat comparable to the size of the disk (some millimeters in diameter), the shadow would contain light and dark fringes around the edge in a circular analog of the Young experiment. Fresnel did not do this calculation, but Poisson did. Poisson also pointed out that – what he was sure would be the *coup de gras* for the wave theory – that the Fresnel theory predicted there would be a bright spot of light in the very center of the dark shadow (far away from the edges). In an *arc de Triomphe*, Poisson whipped out his calculation and slammed it on the desks of the other judges as proof that Fresnel's wave theory was wrong. Naturally, he declared, if you shine light on a circular French franc coin, all you are going to see behind it is a dark circular shadow, and there's

38 I was surprised to learn that the word for a baby shark is a pup.

certainly no way you will see a bright spot at the center of the shadow. Fresnel had to be wrong!

Theoretical physicists often make wild claims like this, and sometimes they are correct. But to be scientifically valid, the assertions need to be tested in the crucible of an experiment. Luckily for Fresnel, the head judge on the prize panel was an experimentalist by the name of Dominique François Jean Arago.[39] Arago set up an experiment that correctly implemented the calculation Poisson made with the theory of Fresnel. Remember, this is all when they are still judging the papers for that prize. A sketch of the experimental set up is shown in Figure 1.9. Here we can see the light source, the circular disk, and the screen with the shadow of the disk, which at the center has that damned bright spot. For this to work, given that visible light is hundreds of nanometers in wavelength, the disk casting the shadow cannot be much bigger than

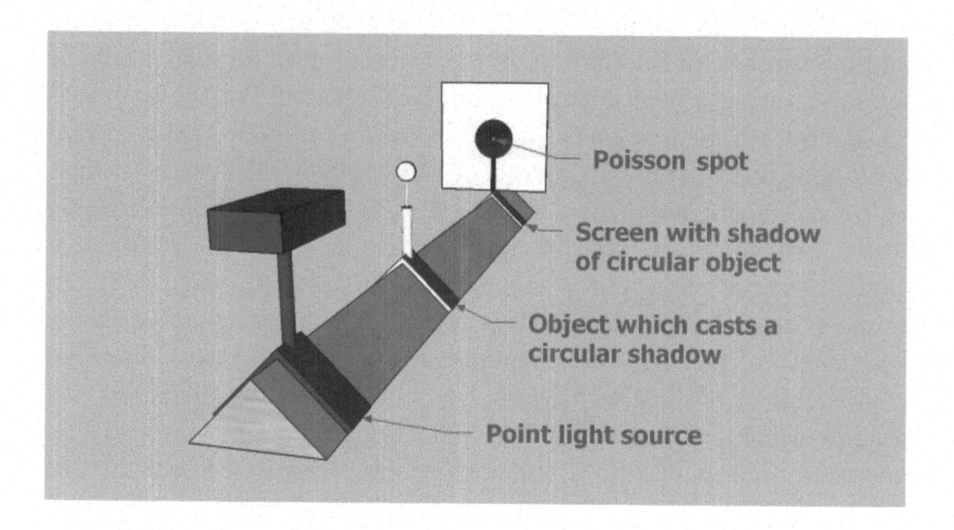

FIGURE 1.9 Arago's setup to look for the Poisson spot. The light source sends out waves that strike the circular disk casting a shadow on the screen. The tiny bright spot in the middle of that shadow is the Poisson spot.[40]

---

39 Arago and the finding of the spot, https://en.wikipedia.org/wiki/Fran%C3%A7ois_Arago.

40 The figure is taken from https://commons.wikimedia.org/wiki/File:Poissonspot_setup_treisinger.jpg. "Arago spot experiment. A point source illuminates a circular object, casting a shadow on a screen. At the shadow's center, a bright spot appears due to diffraction, contradicting the prediction of geometric optics," by Thomas Reisinger. This file is licensed under the Creative Commons Attribution 3.0 Unported license.

that, or the particle theory will work just fine. In typical experiments with visible light, the disk is only a few millimeters in diameter.

Arago carefully aligned all of the optical elements as in Figure 1.9 and duly observed Poisson's predicted bright spot – right in the middle of the shadow – exactly where Poisson argued no spot could be! Arago reported his results to the prize committee, which grudgingly awarded Fresnel the prize for his wave theory of light.[41] After that, scientists abandoned the particle theory, and Poisson had to eat crow (instead of eating Fresnel). These days it is more common to call the Poisson spot the Arago spot or the Fresnel spot so as not to give credit to Poisson – who denied its existence! (However, Arago finally did a proper literature search and discovered the spot had been independently observed nearly a century earlier by two French scientists, Joseph-Nicolas Delisle[42] (in 1715) and Giacomo F. Maraldi[43] (in 1723). So perhaps it should be called the Delisle–Maraldi–Poisson–Fresnel–Arago spot?) Alas, Delisle and Maraldi had no theory to explain their observations at that time, and their citation record on the matter remains spotless. Wherefrom does the mysterious bright spot come? Interference! In Figure 1.10, you can see the dark and bright fringes around the edge of the shadow and the bright spot in the middle. Too faint to see are the bright and dark fringes that move away from the sides and toward the center, but, at the very center, there is constructive interference, and all the bright fringes converge to give that luminous spot. The particle theory had no interference and so no spot.

---

41 *For Arago's report on the finding of the spot, see "Rapport fait par M. Arago à l'Académie des Sciences, au nom de la Commission qui avait été chargée d'examiner les Mémoires envoyés au concours pour le prix de la diffraction," by D. A. F Arago [Report made by Mr. Arago to the Academy of Sciences in the name of the commission, which had been charged with examining the memoirs submitted to the competition for the diffraction prize.]. Annales de Chimie et de Physique. 2nd series (in French).* ***11*** *(1819) pp. 5–30.* From p. 16: *"L'un de vos commissaires, M. Poisson, avait déduit des intégrales rapportées par l'auteur, le résultat singulier que le centre de l'ombre d'un écran circulaire opaque devait, lorsque les rayons y pénétraient sous des incidences peu obliques, être aussi éclairé que si l'écran n'existait pas. Cette conséquence a été soumise à l'épreuve d'une expérience directe, et l'observation a parfaitement confirmé le calcule."* (One of your commissioners, Mr. Poisson, had deduced from the integrals [that had been] reported by the author [i.e., Mr. Fresnel], the strange result that the center of the shadow of an opaque circular screen should – when the [light] rays penetrate it [i.e., the shadow] at slightly oblique incidences – also be illuminated as if the screen didn't exist. This result has been submitted to the test of a direct experiment, and observation has correctly confirmed the calculation.)

42 Delisle's biography, https://en.wikipedia.org/wiki/Joseph-Nicolas_Delisle.

43 Maraldi's biography, https://en.wikipedia.org/wiki/Giacomo_F._Maraldi.

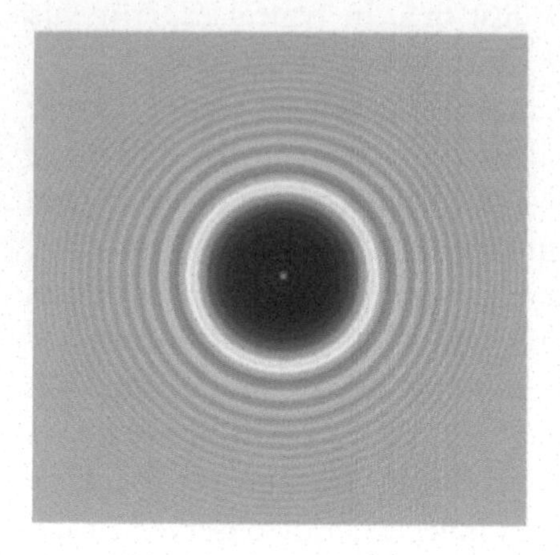

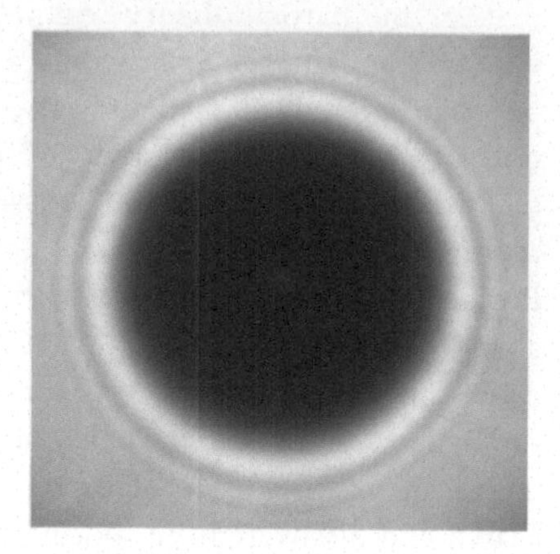

FIGURE 1.10 On the left is a computer simulation of the shadow behind a circular disk using the theory of Fresnel and Poisson. At the center of the shadow is the bright spot. To the right is a photograph of a similar shadow. Although a bit hard to see – if you stare long enough – you can see a dim white spot in the center – which is what Arago observed in his experiment.[44,45]

THE JOURNEY STONE

While Poisson was sputtering, muttering, and puttering with spots, some odd discoveries were taking place in two completely different fields of physics other than optics – the areas of electricity and magnetism. These discoveries would have an enormous impact on the final theory of light. A naturally occurring magnetic material called lodestone had been known to the ancient Greeks and Chinese. (See Figure 1.11.) They noted its ability to attract iron or to attract and repel other lodestones.[46]

44 Figure on the left from https://commons.wikimedia.org/wiki/File:Poissonspot_simulation_d2mm.jpg; "Computer simulation of the Poisson Spot showing in the shadow of a 2-mm-diameter disc at a distance of 1 m from the disc. The point source has a wavelength of 633 nm (e.g., He-Ne Laser) and is located 1 m from the disc," by Thomas Reisinger. This file is licensed under the Creative Commons Attribution-Share Alike 3.0 Unported license.

45 Figure on the right from https://commons.wikimedia.org/wiki/File:A_photograph_of_the_Arago_spot.png. "The shadow of a circular obstacle (5.8 mm) in diameter lit by a ~0.5 mm sunlight source, exposing the Arago spot in the middle of a shadow. Distance from the light source to the obstacle 5 ft. (~153 cm), distance from the obstacle to the paper screen 6 ft. (~183 cm)," by Aleksandr Berdnikov. This file is licensed under the Creative Commons Attribution-Share Alike 4.0 International license.

46 The history of the lodestone, https://en.wikipedia.org/wiki/Lodestone#History.

FIGURE 1.11 A naturally found chunk of lodestone magnetically attracting iron nails to itself.[47]

The Chinese were the first to record, in the 2nd century BC, the use of sharpened lodestone needles floating in water as rudimentary compasses, which shysters used in divination and the alignment of buildings with the north–south direction. However, a grooved magnetic bar – discovered in North America – predates the Chinese writings by a thousand years and suggests the Native American Olmec people had found this compass-like property first. The first recorded use of such compasses for navigation also goes back to China in the 11th century AD. Dry magnetic compasses, with the magnetic needle set on a spindle instead of floating in water, appeared in Europe around the 1300s AD. Even the word lodestone means "leading stone" or "guiding stone" in Middle English.

Apart from the compass, scientists viewed the lodestone as a scientific curiosity. It was something to amuse children and to use in school physics demonstrations. One of the children entertained by a compass

47 Figure taken from https://commons.wikimedia.org/wiki/File%3ALodestone_(black).jpg. "Lodestone," by Teravolt. This file is licensed under the Creative Commons Attribution 3.0 Unported license.

was the young Albert Einstein, whose father gave him a compass and some magnets to play with. Einstein recalls that encountering this toy was a defining moment in his life. From it, he realized that space was filled with mysterious invisible fields that could affect objects at a distance. His general theory of relativity was a culmination of that spark ignited by the toy compass and was to become the last word in classical field theory. But that was to be in the early 1900s. In the 1700s, the lodestone hinted about the existence of a magnetic field. But nobody worried too much about what seemed like a child's toy, and nobody suspected that the magnetic field would have something to do with optics – the study of light.

FRANKLIN, FROGS, AND FRANKENSTEIN

Contrariwise, electricity was all the rage in the 1700s. Another odd phenomenon, static electricity, was also known to the ancients. They observed that on dry days you could pass an ivory comb through your dry hair and get it to attract or repel small bits of paper – again something for children to play with. But unlike the lodestone, many objects could be electrified or de-electrified. A lodestone stayed magnetic all the time. American inventor and diplomat, Benjamin Franklin, was one of the first to systematically study electrified objects by experimenting with rods of rubber and glass. He electrified these things by rubbing them with felt or cat hair. He concluded that electricity was something like a fluid with positive charges that could move around from object to object and with negative charges that stayed put. (He got this precisely backward – his biggest blunder. The negative charges move, and the positive charges remain put. This mistake confuses my students to this day.) From his observations, he posited that like charges repel, and opposite charges attract.[48] Franklin was the first to propose an experiment to show that lightning was an electrical phenomenon. In his famous test, carried out in 1752, he flew a kite into a thunderstorm and observed that the string (attached to a metal house key) electrified a very early prototype of a voltmeter.

This experiment was perilous since the string could have acted as a type of lightning rod – as Georg Wilhelm Richmann in Saint Petersburg sadly discovered in August 1753 when, while trying to replicate the experiment, he was struck by lightning and killed. Franklin's studies of

48 More on Franklin, https://en.wikipedia.org/wiki/Benjamin_Franklin.

the relationship of lightning to electricity inspired his invention of the sharpened metal lightning rod that soon found its way onto tall buildings all across the country.[49] (There are some earlier claims of lightning rods installed on churches in Europe. These include a metal ball with spikes sticking out of it placed atop the Leaning Tower of Nevyansk, and also a meteorological machine devised by a Premonstratensian Priest of the name Prokop Diviš, which was erected in the town of Přímětice, near Znojmo, Moravia, in June of 1754.)

Franklin's law of charges was quantified theoretically and experimentally by the French scientist, Charles-Augustin de Coulomb. The law is now called Coulomb's law – electric charges produce electric force fields that repel or attract other charges. The bits of paper jumping onto the ivory combs of the Ancient Greeks were simply manifesting this law.

Picking up where Franklin left off was the Italian scientist Alessandro Giuseppe Antonio Anastasio Volta. Volta, for whom we named the household unit of volts, was the inventor of the electric battery. This device consists of electrochemical cells in a pile composed of salt and acid, like in your car battery. This development was a revolutionary step in the study of electricity. Instead of waiting for a dry day to rub a rubber rod with a cat, or a wet day to fly kites in a thunderstorm, the battery delivered controlled amounts of electrical current at the flip of a switch.[50] Volta was inspired to invent the battery to disprove a hypothesis of his competitor, Luigi Galvani, who had been experimenting with electricity and twitching frog legs. Galvani observed that the frog leg was a source of electricity and concluded that all electricity came from living things. (What about lightning?) Volta's battery disproved Galvani's theory – a battery is not alive. Nevertheless, the notion that electricity was associated with living things inspired Mary Wollstonecraft Shelley to write her famous novel *Frankenstein*, where she replaced the frog legs with human body parts. Indeed, you can attach a battery to a dead frog's severed limb, and watch it twitch as the electrical current induces contractions in the muscles.

ØRSTED UNLODES

After that, the battery became a must-have device in physics classroom demonstrations. Teachers could twitch the legs of frogs, make sparks,

49 More on Lightning Rods, https://en.wikipedia.org/wiki/Lightning_rod#Russia.
50 Biography of Volta, https://en.wikipedia.org/wiki/Alessandro_Volta.

electrocute themselves, and do all sorts of other cool things.[51] In one such demo, serendipity found the Danish scientist Hans Christian Ørsted.[52] Often people think that, when a scientist makes a significant discovery, they yell eureka! However, in reality, they are much more likely to yell – what the *hell* was *that!?* Ørsted's finding was of the latter kind. Ørsted was an early fan of Volta's batteries and built a number of them for himself to carry out his investigations into electricity. In one of his lecture-demonstrations, he happened to have a compass on the same trolley with his battery and current-carrying wire. If you have ever taken an undergraduate or high school physics course, these demos consist of a rolling cart with physics gee-gaws on them used to hammer home the point of the lecture. In 1820, while giving such a demo, Ørsted noticed that whenever he threw the switch to the battery to let electrical current flow through a wire, the nearby compass needle would rotate one way, and when he turned the switch off, the needle would rotate back. Did he yell eureka? No. Quietly he thought to himself – what the *hell* was *that!?* (See Figure 1.12.)

Ørsted said nothing but went back to his lab and, in secret, carried out a series of experiments with compass needles and current-carrying wires. He established conclusively for the first time that there was a connection between electricity and magnetism. The current in the wire produced a magnetic field that caused the compass needle to turn. The magnetic field encircled the wire in loops, and the strength of it fell off inversely with distance. I cannot overstate how important this discovery was – two seemingly distinct realms of physics – thought to be unrelated for thousands of years – were suddenly seen to be connected. This discovery is the first step to what physicists call a unified field theory, where two seemingly unrelated subjects are suddenly now perceived to be part of a unified whole. The distinct areas of research on electricity and

51 While conducting such a demo in my class of 70 engineering students, I was demonstrating how a high-frequency AC electromagnetic device called a Tesla coil could shoot sparks up to 3 m and make nearby long tubular fluorescent light bulbs glow, even if the bulbs were not in their sockets. To avoid electrocution, I attached one end of a wire to the metal prongs on the lamp and plugged the other end into a nearby wall electrical outlet. The idea was that if the lightning struck the bulb, the electricity would go through the wire into the ground. Unfortunately, that particular wall socket was disconnected, and so when the lightning struck – it traveled through me! To the amusement of my students, I yelled a very bad word and flung the bulb into the wall where it shattered to smithereens. Even more amusing was that the bulb contains mercury, and so the students got out of class early while the guys in the hazmat suits showed up to clean out the room. I'm not allowed to use the Tesla coil anymore. You can find out about the Tesla coil here https://en.wikipedia.org/wiki/Tesla_coil.

52 Biography of Ørsted, https://en.wikipedia.org/wiki/Hans_Christian_%C3%98rsted.

FIGURE 1.12 Artist's depiction of Ørsted's discovery that a current-carrying wire (in his hands) causes a compass needle to align perpendicular to the wire. The thing that looks like a blue cocktail (that you might serve at an Andorian wedding) is part of a battery.[53]

magnetism are now called one thing – electromagnetism. In 1854, Ørsted published his findings in a book with the spiritual title, *The Ghost in Nature*. It was a breakthrough.

The findings of Ørsted were phenomenological. He reported his experimental results in detail, but he had to await the theorists to come and put these data into a simple mathematical framework. A pair of French experimentalists, Jean-Baptiste Biot and Félix Savart, carried out similar experiments to Ørsted's, but also managed to quantify the magnetic field around the wires in what is called the Biot-Savart law.[54] Another French scientist, André-Marie Ampère, for whom

53 Figure from https://commons.wikimedia.org/wiki/File:Oersted_discovers_electromagnetism.jpg. "Ørsted discovers electromagnetism in 1820. Engraving with later coloration," by an unknown author. This work is in the public domain in its country of origin and other countries and areas where the copyright term is the author's life plus 70 years or less.

54 More on the Biot-Savart law for magnetic fields around a wire, https://en.wikipedia.org/wiki/Biot%E2%80%93Savart_law.

the unit of electrical current the amp is named, completed the theory. The production of magnetic forces and fields by current-carrying wires was now set in stone. It was also unset in lodestone. Before Ørsted's discovery, the only known way to produce magnetic fields was with naturally occurring permanent magnets like magnetite. Now you could generate magnetic fields, literally, with the flip of a switch. A favorite device, a tube of coiled wire called a solenoid, also invented by Ampere, could produce robust and uniform magnetic fields inside the coil, with a strength proportional to the magnitude of the current and the number of loops. So goodbye lodestones – with a battery and a coil of wire, you could produce reliable and adjustable magnetic fields from electricity. What next?

ELECTRIC BOY REWINDS

The person singlehandedly responsible for developing the ideas of gravitational, electric, and magnetic fields in space, was the self-made English scientist, Michael Faraday. You can find the story of his life told in the episode "The Electric Boy" on the newly minted television series *Cosmos*, which is narrated by Neil DeGrasse Tyson.[55] Faraday had the equivalent of elementary school education, and in his teens was apprenticed to a bookbinder. From that humble start, he went on to become a Fellow of the Royal Society of London and one of the most influential physicists of all time – right up there with Newton. Einstein kept a portrait of Faraday next to that of Newton on his office wall. Faraday did not know much math, but he had a vivid imagination, and he could see in his mind's eye the invisible fields that would someday so shock Einstein as a child. Faraday, between hours of binding books together, set up a lab in the bookbinder's shop and began carrying out experiments that would change the course of science. These experiments led directly to the unification of optics, magnetism, and electricity.

Ampere's solenoids particularly enthralled Faraday, those tube-shaped coils of wire, which, when attached to the poles of Volta's battery, would conduct a current that would, in turn, produce a powerful magnetic field. Faraday had this idea that if electricity could produce magnetism, then by symmetry, magnetism should generate

55 Synopsis of the *Cosmos* episode "Electric Boy," on the life of Faraday is at https://en.wikipedia.org/wiki/The_Electric_Boy. (You should watch it!)

electricity. To make a solenoid, you take a cardboard tube and wrap the outside round and round with insulated conducting wire. (I made one for my third-grade science project that was strong enough to lift a 20-kilogram block of iron.) You hook the two ends of the wire to the poles of a good battery, and as the current flows, the tube acts like a bar magnet with one end the South Pole and the other end the North Pole. The strength of the field is proportional to the current and the number of windings. In Faraday's day, insulated copper wire was difficult or expensive to come by. He would take the bare copper wire and carefully wrap the strands with glued strips of felt insulation – the felt which he had deliberately cut off the back of his wife's evening gowns! She was not able to tell for some years due to the way he hung them back in the closet.[56] (The Faradays did not get out much.) The more windings you put around the tube, for a fixed current, the stronger the magnetic field. (For my science fair project, I attached the tube to the spindle of an old mechanical sewing machine and spun the wire onto it while working the pedal. That way, I could quickly get a huge number of loops and a humongous magnetic field.)

In 1831, Faraday had the idea to revise the experiment to see if magnetism could produce electricity – the reverse of what Ørsted had found. Faraday disconnected the solenoid wires from the battery and hooked them up to an ammeter, which is a device for measuring electrical current. He took a lodestone bar magnet and carefully – very, very, very slowly – lowered it down into the hollow tube of the solenoid, looking for any sign that the magnetism of the bar magnet would induce a current in the wire. Faraday saw nothing. Very, very, very slowly, he removed the bar magnet, and went back to his wife's closet for more felt. He then wrapped a few hundred more coils of wire around the tube, slowly lowered the bar magnet into the container – and again saw nothing. Faraday repeated this several times and saw nothing several times. Frustrated, he grabbed the bar magnet and yanked it quickly out of the solenoid. Then – and only then – the ammeter needle suddenly swung around and pegged – showing a burst of maximum current. Then the needle went back to zero when Faraday stopped yanking on the bar. In one of the greatest eureka moments of history, Faraday exclaimed – what the

56 Hirshfeld, Alan W. *The Electric Life of Michael Faraday*. New York: Walker & Co., 2006. www.worldcat.org/oclc/907144505.

hell was *that!?* A *stationary* magnetic field did not produce an electric current, but a *moving* magnetic field did! This observation was a stunning result and nearly closed the theoretical loop opened by Ørsted. A *steady* current flow produces a magnetic field, and a *moving* magnetic field creates a current. We call this effect Faraday's law of induction since the moving magnetic field *induces* a current in the wire.[57]

This result meant you could produce electric currents - not only with acid-leaking batteries - but simply by moving a magnet in and out of a coil of wire. You convert mechanical work done by your arm into an electromotive force - which is the electric voltage that pushes current in a cable. That observation is the principle behind an electric generator. The configuration is a bit different, but by turning a crank connected to a coil of wire immersed in a magnetic field, you could produce either alternating current or direct current.[58] This discovery was the first step in building the electric power grid.[59] We show an improved version of Faraday's experiment in Figure 1.13. We replace the bar magnet with a primary solenoid that is attached to a battery. When the current flows, the primary acts as a bar magnet, which we then push and pull in out of the secondary solenoid to produce a current. The current is forced down the wire by an electromagnetic field. The changing magnetic field creates an electric field in the secondary. That is what drives the current in the secondary. Finally, if we replace the battery with an alternating current generator and place the primary inside the secondary, the primary produces an alternating current, which creates a new current in the secondary. This process occurs even if the primary coil does not touch the secondary coil and is the principle behind the alternating current transformer - another critical element of Tesla's design for the power grid.[60] By changing the number of loops in the secondary versus the

---

57 Faraday's Law of Induction, https://en.wikipedia.org/wiki/Faraday%27s_law_of_induction.

58 Overview of the electric generator, https://en.wikipedia.org/wiki/Electric_generator.

59 Around 1900, Edison and Tesla were in a completion to create an electrical network in New York State that would be powered by generators at Niagara Falls. Edison's proposal used direct current and would have powered Buffalo, NY. Tesla's proposal applied alternating current and would power the entire state, including New York City. Tesla won. For more on the War of The Currents, see https://en.wikipedia.org/wiki/War_of_the_currents. A film on this story just came out, titled The Current War, www.imdb.com/title/tt2140507/.

60 See https://en.wikipedia.org/wiki/Transformer.

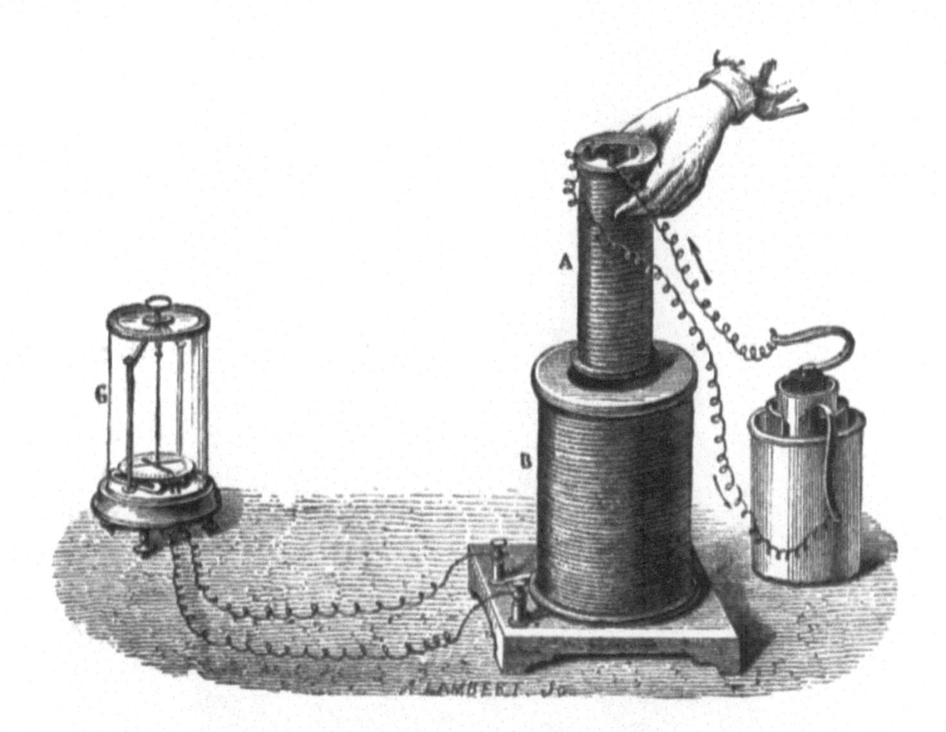

FIGURE 1.13 A modified version of the Faraday experiment. We replace the bar magnet with a primary coil of wire (A) attached to the battery on the right. The primary coil is a solenoid that produces a magnetic field like a bar magnet. If we drop the primary coil into the secondary coil (B), a current is induced in the secondary coil's wire and measured at the galvanometer (G). (A galvanometer is an old name for a current-measuring device that is now called an ammeter. We name it the galvanometer after Luigi Galvani, whose electric experiments with jumping frogs led to the writing of Frankenstein.)

primary, you can use Faraday's law to increase or decrease the voltage – taking the thousands of volts in a power line down to a hundred volts in your wall socket. The take-home message is that a changing magnetic field produces a changing electric field – even through empty space.

I use Faraday's law to charge my *Sonicare Series 2 Plaque Control Rechargeable Toothbrush*. Unlike my rechargeable razor, which has two metal prongs through which the current flows to charge the battery, the

FIGURE 1.14 Michael Faraday is holding a bar of glass he used to demonstrate the Faraday effect – that a magnetic field applied to the glass would change the polarization of light passing through the glass.[61]

toothbrush has no exposed metal parts at all.[62] The cylindrical plastic base of the toothbrush has a circular hole that fits over a tubular stub on the charging stand. The stub has a primary coil with an alternating current in it under the plastic. The toothbrush has a secondary coil, attached to a rechargeable battery, under the plastic in the handle. The

61 From https://commons.wikimedia.org/wiki/File:Faraday_photograph_ii.jpg, "Image of Michael Faraday holding a glass bar of the type he used to show that magnetism affects light; part of a photograph by Maull & Polyblank." This work is in the public domain in its country of origin and other countries and areas where the copyright term is the author's life plus 70 years or less. (For years I thought this was an image of Faraday holding either a solenoid or a cigar.)

62 For the theory of my toothbrush, see https://en.wikipedia.org/wiki/Sonicare. (The fact that my toothbrush has its own Wikipedia page explaining all this is fantastic. I wish I had found this page before taking my toothbrush apart to see how it worked. Instead, I irreparably damaged it and had to buy a new one.)

primary in the stub sets up a changing magnetic field, which sets up a changing electric field in the toothbrush, which produces the current that charges the battery. The fields move through space – and this is the crucial point for our story here – the charger and the toothbrush do not even need to touch. Faraday's law of induction tells us that changing magnetic fields will induce electric fields in space. Franklin's and Coulomb's laws tell us that electrical charges will produce electric fields in the vacuum of space. Ampere's law tells us currents induce magnetic fields in space. There was one missing law needed to close the loop and to show light itself was an electromagnetic wave. (Yes, there is a point to this whole story.)

In addition to the law of induction, Faraday made another conjecture that magnetic fields could affect the polarization of light – since he had a hunch that electricity, magnetism, and optics were all somehow connected. Indeed, Faraday did experiments on light passing through bars of glass that were stuck inside his solenoids and showed the polarization angle rotated as a function of the strength of the magnetic field. This rotation is called the Faraday effect and is used to this day to make elaborate optical and laser systems.[63] Faraday was very close. He had a hunch that light, electricity, and magnetism were all related, but it was only a hunch.

We have set the stage for the grand unification of electricity, magnetism, and optics. Faraday had laid the groundwork, and his vision of electric and magnetic fields in space allowed us to think of these fields independently of the electrical charges and currents that produced them. There was one missing law to close the last loophole, and then all that remained to do was to put the rules into a robust mathematical framework. Alas, with his sixth-grade education, the tremendous visionary Faraday, in the words of Neil DeGrasse Tyson, "Could not do the math." In 1867, at age 75, Faraday died peacefully at his home from Alzheimer's disease. By then, a person who *could* do the math was at the peak of his calculational prowess, and this Mathematical Maestro was Maxwell.

ALL'S WELL THAT'S MAXWELL

The Scotsman, James Clerk Maxwell, was not an experimentalist, but he was a keen mathematical physicist.[64] When Faraday was fading, Maxwell

63 The Faraday effect uses magnetic fields to change the polarization of light. See https://en.wikipedia.org/wiki/Faraday_effect. It suggests a connection between electromagnetism and optics.

64 You can find the Wikipedia biography of Maxwell here https://en.wikipedia.org/wiki/James_Clerk_Maxwell. For reasons the Scotts only know, we pronounce his middle name "Clerk" instead of "Clark." Maxwell was an old-school Sheldon-like character (from the televisions series *The Big*

was at the height of his career, having had trained in mathematics and physics at the University of Cambridge. He eventually held the professorship Chair of Natural Philosophy at King's College, London, where he did much of his famous work, including the unification of the theories of light, electricity, and magnetism. Maxwell would regularly attend lectures by Faraday, who was 40 years older, and by that time already showing signs of dementia. However, Maxwell took enough away from Faraday's lectures and writings to carry the electromagnetic baton over the optical finish line.

The first thing Maxwell did was to put all the known laws of electricity and magnetism into a single unified set of four mathematical formulas called Maxwell's equations. It is indeed a bit of a snub to the rest of the community because "Maxwell's" equations contain Coulomb's law, Ampere's law, and Faraday's law of induction.[65] It was Faraday's law of induction that puzzled Maxwell considerably. Being a theorist, he could abstract to a situation where you would be far out in space where there were no charges and no currents. However, you still could have a changing magnetic field inducing an electric field without anything in between – just like with my toothbrush.

Here Maxwell made his most significant contribution to electromagnetism by just thinking about things and without even going into the lab. He conjectured that, by symmetry, if a changing magnetic field could induce an electric field, then a changing electric field should also produce a magnetic field. By golly – *that* was a winner! Maxwell conducted a thought experiment involving two wires – one attached to the terminals of a battery and the other connected to the two plates of a metal capacitor (Figure 1.15). (A two-plate capacitor is a device that stores electric energy in the form of an electric field between the plates.[66]) Maxwell reasoned as follows. When you hook a capacitor up to a battery, the battery charges the

---

Bang Theory). When he was an undergraduate at Cambridge University, he would run up and down the hallways of his dormitory in the middle of the night for exercise. His clunky Scottish shoes made a terrible racket. He did this every night until his classmates (who called him "Dafty") threatened to beat him up. See, Mahon, Basil. *The Man Who Changed Everything: The Life of James Clerk Maxwell* . Chichester, West Sussex, England: Wiley, 2003. Ch.1, www.worldcat.org/oclc/52358254.

65 There are a couple of Gauss' laws that I left out, but they do not contribute to the point of this discussion. For a complete and gory overview of all of Maxwell's equations see https://en.wikipedia.org/wiki/Maxwell%27s_equations. For a more historical point of view see https://en.wikipedia.org/wiki/History_of_Maxwell%27s_equations.

66 For an overview of electrical capacitors, see https://en.wikipedia.org/wiki/Capacitor. For an overview of flux capacitors see https://en.wikipedia.org/wiki/DeLorean_time_machine#Flux_capacitor. (Flux capacitors do not exist.)

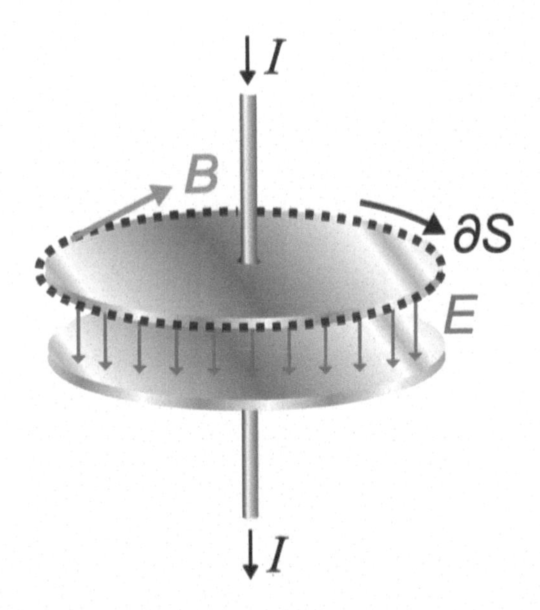

FIGURE 1.15 A schematic of Maxwell's displacement-current experiment. If the current I, produces a magnetic field B around itself, then so must the changing electric field E, provide a magnetic field about itself.[67]

plates of the capacitor. As the current flows from the batteries into the plates, it produces a changing electric field that increases until the capacitor is fully charged and then the current stops. While the current is flowing, and the electric field between the plates is changing, the current produces a magnetic field in circular loops around the wires via Ampere's law.

What happens around the changing electric field between the plates? Does the magnetic field go suddenly to zero when the wires hit the plates? Maxwell could not stand for that, as it would mean a weird discontinuity in the magnetic field at the plate boundary. He hypothesized the changing electric field between the plates produced a magnetic field outside the plates in the same shape and strength as the magnetic field outside the wires. With about three lines of algebra, he worked out what is called Maxwell's displacement law – a changing electric field produces a magnetic field – and thereby, he discovered the missing link

67 Figure from https://commons.wikimedia.org/wiki/File:Displacement_current_in_capacitor_1.svg. "Displacement current in a capacitor with line integral designed to find the magnetic field at the edge of the capacitor." The multiple authors have released the figure into the public domain for any use whatsoever.

that closed the logical loophole in Maxwell's equations. Maxwell called the changing electric field a displacement current, as it behaved in almost every way, like a real current moving between the plates.

We now have Faraday's law of induction – a changing magnetic field produces an electric field in space. We also have Maxwell's law of displacement current – a changing electric field generates a magnetic field in space. That was it – the electric and magnetic fields formed a *positive* feedback loop. The electric field produces a magnetic field, which creates an electric field, which provides a magnetic field, and so on – forever. Maxwell predicted electromagnetic oscillations could propagate in space – and that they were a wave. But what kind of wave? Well, you should be able to guess by now – a light wave.[68]

FIAT LUX!

In one remaining flourish of mathematical rigor, Maxwell proved that electromagnetic oscillations in space were light waves. In Maxwell's equations in empty space, absent of currents or charges, the number of relevant equations reduces to just two – Faraday's law of induction, and Maxwell's displacement current law. Even far from any currents or sources the, equations contained relics of the other rules. That is, they included two fundamental physical constants, ε_0 and μ_0. (They are pronounced epsilon-naught and mu-naught, respectively. Since ε is the Greek symbol for the letter "e" and μ is the Greek symbol for the letter "m", they mean electric and magnetic, respectively.) Initially, ε_0 was a constant of proportionality that told you how much electric field was produced by a given charge. The μ_0 was a constant of proportionality that told you how much magnetic field was provided by a given current.

What was odd, to Maxwell at least, was that two equations – Faraday's law and Maxwell's law – survived in empty space far away from any currents or charges but still had these two constants floating around. Hot on the trail of electromagnetic oscillations in space, Maxwell combined the two laws into a single equation, and low and behold the result was a wave equation. Scientists had studied wave equations for over a hundred years before Maxwell, such as in water waves, sound waves, and so on, and so when you see a wave equation, you immediately recognize it as such. But what kind of wave? A wave

68 For more on Maxwell's displacement current, see https://en.wikipedia.org/wiki/Displacement_current.

equation contains only one single constant c, which is the velocity at which the wave moves. Staring at his wave equation, Maxwell could immediately read off the value of c, and it was inversely proportional to the square root of the product of the electric and magnetic constants. The magnetic constant had been measured very carefully in the experiments of Ampere. The electric constant had been measured very accurately in the analyses of Coulomb. These two constants, magnetic and electric in origin, were thought to have nothing to do with each other – much less with light waves. However, physicists had determined their values so well, that Maxwell could plug them into his formula and estimate the velocity of the electromagnetic wave. He found that it was $c = 300{,}000$ km/s. That is pretty darn fast. What the heck moves that fast? By the mid-1800s – everybody knew! – it was light. (See Figure 1.16.)

In 1679, the not-burned-at-the-stake Italian scientist, Galileo Galilei, carried out the first attempt to measure the speed of light.[69] When not dropping his massive balls off the leaning tower of Pisa, he proposed and carried out experiments on the hilltops of Pisa, where he and an assistant would open and close the shutters on lanterns on different hills, and try to time the speed of light. He-who-always-goes-by-his-first -name Galileo concluded that, if not instantaneous, light moved faster than his experiment could measure.

The first quantitative experimental measurement was an astronomical one. Since light moves speedily, you need to measure its velocity over vast distances to get an accurate result. By the mid-1600s, the orbits of the moons of Jupiter were so well known that astronomers could predict the moment they would appear or disappear from behind Jupiter's disk – down to a minute or so. It was like clockwork. However, the Danish astronomer, Ole Rømer, found the clocks were running a bit strangely. Sometimes the moons of Jupiter would move into view a few minutes earlier than expected and sometimes they would appear a few minutes later than expected. What was wrong with these celestial clocks? Well, nothing. Rømer noticed the moons appeared early when the earth was on the same side of the sun as Jupiter and appeared late when the earth was on the opposite side of the sun. Rømer conjectured that the puzzling behavior of the clockwork moons was due to the speed

69 Biography of Galileo can be found here https://en.wikipedia.org/wiki/Galileo_Galilei. Queen's Bohemian Rhapsody, featuring the refrain, "Galileo! Galileo!" can be found here www.youtube.com/watch?v=fJ9rUzIMcZQ.

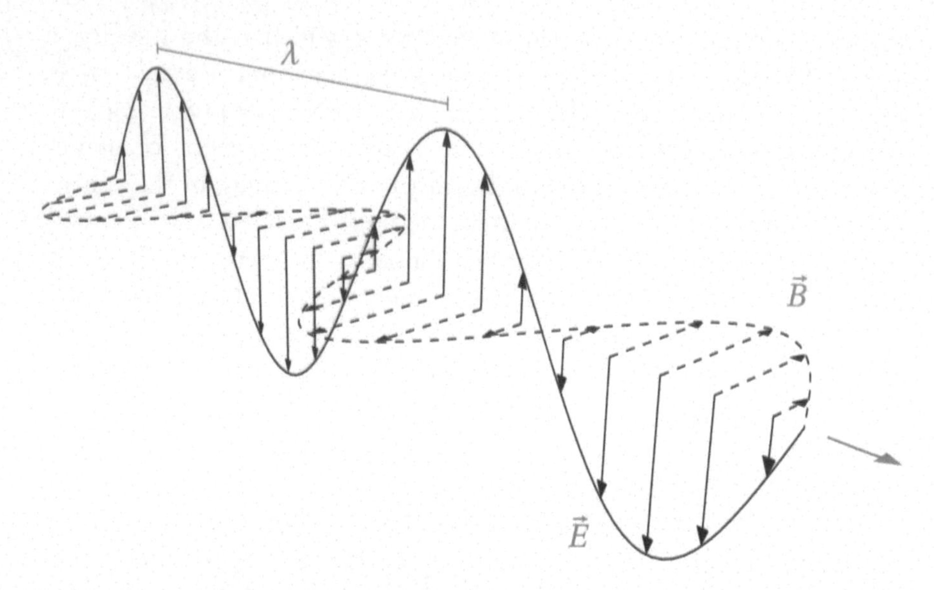

FIGURE 1.16 Maxwell's electromagnetic wave. A changing magnetic field $\vec{B}$ induces a changing electric field $\vec{E}$ that produces a changing magnetic field $\vec{B}$ in a positive feedback loop that can propagate billions of lightyears in space. We can explain all polarization experiments from the observation that the electric field is always polarized (or perpendicular) to the direction of movement.[70]

of light. Knowing the diameter of the earth's orbit around the sun, and by timing the delays and advances of the appearances of the moons, he estimated the speed of light to be 220,000 km/s, which is about 26% below the now accepted value of 300,000 km/s. (It takes light about 17 minutes to cross the diameter of the earth's orbit.) His answer was off due to crappy clocks and a lack of precise knowledge of the diameter of the earth's orbit. (Rømer's boss, the director of the Paris Observatory, did not buy this interpretation, but *who* remembers *his* name?)

By Maxwell's time, earth-based optics experiments had determined the value to be 300,000 km/s – precisely the speed that Maxwell found hidden in his new theory.[71] Here we have the grand unification of electricity, magnetism, and light. Maxwell combined these three separate

70 Figure from https://commons.wikimedia.org/wiki/File:Electromagnetic_wave.svg. "Electromagnetic wave," by Lenny Wikidata. This file is licensed under the Creative Commons Attribution-Share Alike 3.0 Unported license.

71 For more history on the speed of light, https://en.wikipedia.org/wiki/Speed_of_light#Connections_with_electromagnetism.

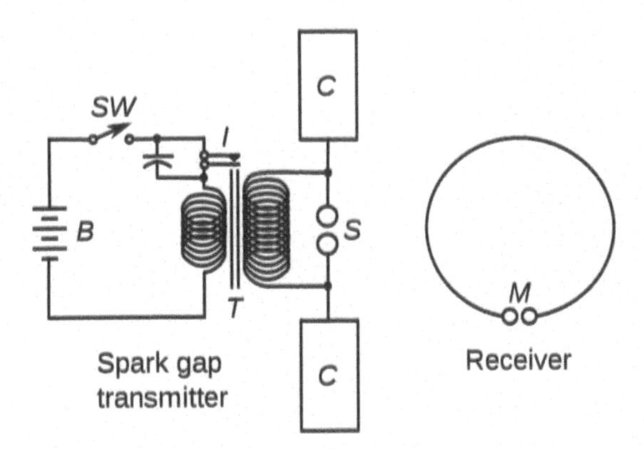

FIGURE 1.17 "Experimental circuit used by Heinrich Hertz in 1887 to discover radio waves. It is a spark gap radio transmitter (left) consisting of a dipole antenna made of two horizontal wires with metal plates on the ends (C) to add capacitance, with a spark gap (S) between them, attached to an induction coil (T) powered by a battery (B). Pulses of a high voltage are applied to the antenna by the induction coil and cause sparks across the spark gap, which excite standing waves of current in the antenna, causing it to radiate electromagnetic waves (radio waves). The waves were detected by a crude receiver consisting of a resonant loop antenna (right) made of a circle of wire, with a micrometer spark gap (M) between its ends. The device produced single short pulses of radio waves; when Hertz pushed the switch (SW) in the primary circuit of the coil, a single spark would jump across the transmitting antenna, creating a radio wave pulse that would induce a single tiny spark in the receiver loop. The frequency of the waves was determined by the length of the antenna, which acted as a half-wave dipole; the short antennas Hertz used produced high-frequency waves in the UHF band, about the frequency of television transmitters."[72]

areas of research with just his four equations. Even better, we could use Maxwell's theory to derive the Huygens–Fresnel wave theory, as well as the wave theory of rainbows and polarization birefringence in Icelandic spar, thus capping the connection with the earlier wave theory of light. However, it predicted something extraordinary – that visible light was just a small band of the electromagnetic waves – there were all sorts of other forms of light that could not be seen, such as radio waves and X-rays. His peers found *this* hard to believe.

72 From https://commons.wikimedia.org/wiki/File:Hertz_transmitter_and_receiver_-_English.svg. "Hertz transmitter and receiver" by multiple authors. This file is licensed under the Creative Commons Attribution-Share Alike 3.0.

WATT IS LOVE? DON'T HERTZ ME, NO MORSE!

The German physicist, Rudolf Heinrich Hertz, took up the experimental gauntlet where Maxwell's theory had left off. (It was he for whom the unit hertz is named, such as this computer has a 20-gigahertz processor speed. Hertz, the unit, means cycles or oscillations per second, and we're talking about the number-of-oscillations per second in the light waves.) As we noted above, the cognoscenti did not take kindly to the idea that there was all this electromagnetic radiation – light waves – outside of the visible spectrum, which we could not see with our own eyes.

Scientists had hitherto discovered invisible ultraviolet light just beyond the purple end of the spectrum, and heat radiation or infrared radiation just beyond the red end of the spectrum. We can see neither of these with the naked eye, and these two had wavelengths (peak-to-peak separation of the wave fronts) on the order of hundreds of nanometers. To believe in them was a small leap of faith. But if Maxwell was right, there were waves (now called radio waves) out there with wavelengths of meters or even kilometers that nobody had ever seen and that nobody had ever guessed were there. Who could believe that? Well, Hertz did.[73]

During 1886–1889, Hertz stumbled upon a way to make a radio transmitter and receiver while he was investigating the charging of metal spheres (Figure 1.17). Two spheres were attached to a long wire that had a gap in the middle. As the spheres charged up, they sparked across this gap so that it gave off high-frequency radio waves at 50 megahertz, which corresponds to an electromagnetic wavelength of 6 m (about the frequency of a television station). Hertz set up a similar device to use as a receiving antenna about 12 m away and saw induced sparking on that device, giving evidence the electromagnetic radio wave had moved from the sender to the receiver. These experiments are widely seen as the first demonstration of radio waves and as verification of Maxwell's equations. When asked about the importance of his tests, Hertz replied, "It's of no use whatsoever [...] this is just an experiment that proves Maestro Maxwell was right – we have these mysterious electromagnetic waves that we cannot see with the naked eye. But they are there." When asked about the applications of his work, Hertz stated, "Nothing, I guess."[74] We should compare this response to

73 A biography of Hertz can be found here https://en.wikipedia.org/wiki/Heinrich_Hertz.

74 From the Institute of Chemistry, Hebrew University of Jerusalem, in the collected papers of Heinrich Rudolf Hertz.

that of Faraday while he was giving a demonstration to the British Parliament. When a Minister of Parliament demanded to know, "Of what use is this electricity!?" Faraday wisely replied, "Minister, I do not know, but I assure you that someday you will *tax* it!"

MARCONI AND CHUTZPAH

As usual, there were complaints from the electromagnetic intelligentsia. Several physicists pointed out that at a wavelength of 6 m, with a separation between the antenna and receiver of 12 m, Hertz had not demonstrated electromagnetic waves at all but just confirmed Faraday's law of induction. The objection was that Hertz's experiment was less like a radio and more like a garage-sized mockup of my toothbrush. Somebody had to experiment with much larger distances to convince the critics otherwise. At times like these, "When the going gets weird, the weird turn pro."[75] And so we encounter a crazy Italian by the name of Guglielmo Giovanni Maria Marconi.[76] Marconi was more of a businessman than a physicist, which is perhaps why the physics community did not take him seriously at first. Building on the work of Hertz and others, he first hit a breakthrough when, by playing around with radio transmitters and receivers, he was able to produce an improved system that could transmit radio waves outdoors over a distance of 3.4 km – even over small hills. He kept increasing the range and improving the system until he was able to transmit telegraphy signals wirelessly up to hundreds of kilometers. This result should have put the critics of Hertz to rest – since that distance was much longer than a wavelength. Alas, the weak Morse-code signals were not all that consistent or convincing. It was then that Marconi went utterly bonkers.

In 1901, with funding from the British government (which was interested in ship-to-ship communications),[77] he set up a transatlantic

75 Quote by Hunter S. Thompson, "Fear and Loathing at the Super Bowl" (*Rolling Stone* #155) (28 February 1974); republished in *Gonzo Papers, Vol. 1: The Great Shark Hunt: Strange Tales from a Strange Time* (1979), p. 49.

76 For a biography of Marconi see https://en.wikipedia.org/wiki/Guglielmo_Marconi.

77 Marconi's invention caught the public's eye for the role it played in maritime rescues, such as with the sinking of the *Titanic*. See www.theatlantic.com/technology/archive/2012/04/the-technology-that-allowed-the-titanic-survivors-to-survive/255848/.

system with a transmitter in Cornwall, UK and a receiver in Nova Scotia, Canada. The critics scoffed. Marconi! The earth is round! Your radio waves move in a straight line! Your radio signals will go off into space and never return! What the critics did not know, and Marconi did not know either, is that the earth is surrounded by a spherical atmospheric layer called the ionosphere. While this layer is transparent in the visible part of the electromagnetic spectrum – invisible to the naked eye – it acts as a conducting metal shell for radio waves. The wave was launched up from Cornwall, bounced off the ionosphere, and ricocheted back down to Nova Scotia, where the receiver recorded the dots and dashes of Marconi's transmitted Morse code. The crazy Italian had demonstrated that Maxwell's radio waves were able to propagate over the Atlantic Ocean, and he discovered the ionosphere to boot. Crazy like a fox! In 1909 Marconi received the Nobel Prize in physics for his work on radio waves. Maxwell, long dead, had finally and unambiguously been vindicated.

MAXWELL'S RAINBOW

We wind up chapter one with the following observations. For 150 years – from Newton and Huygens until Hertz and Marconi – scientists argued about the true nature of light. Newton's particle theory held sway for the first 100 years, but then Young's two-slit experiment revitalized the Huygens wave theory into the Huygens–Fresnel wave theory. The observation of the Poisson-Arago spot overturned the particle theory in favor of the new and improved wave theory. Maxwell refined the Huygens–Fresnel approach with his equations of electromagnetism that, as a freebie, predicted that light was an electromagnetic wave. With the method of Maxwell, one could thoroughly explain the double rainbow, the double refraction of polarized light through Icelandic spar, and the Poisson–Arago spot.

Finally, as a real shocker, Maxwell predicted there were invisible electromagnetic waves with wavelengths much longer or much shorter than those of visible optical waves. (See Figure 1.18.) Hertz and Marconi confirmed the existence of the longer radio waves, and around the same time, 1895, Wilhelm Conrad Röntgen discovered the invisible

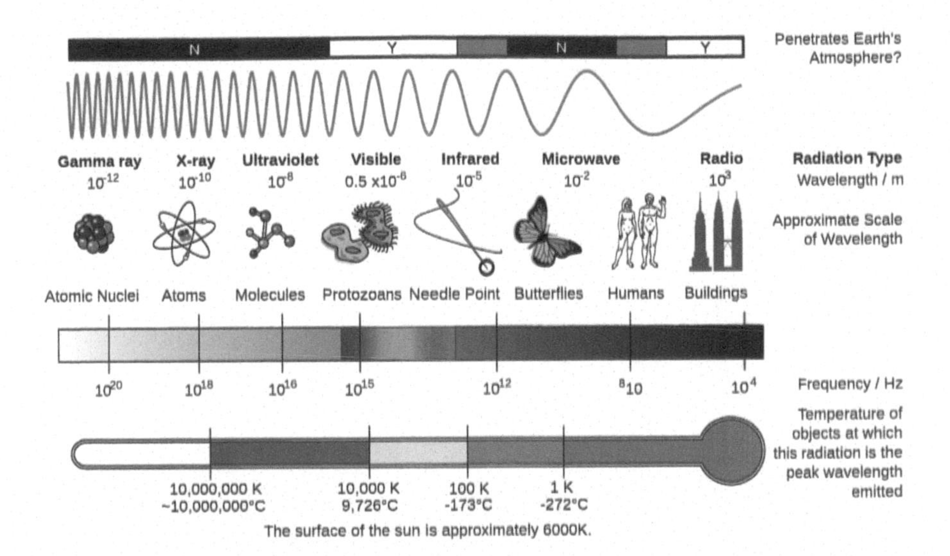

FIGURE 1.18 Maxwell's rainbow. The universe is filled with invisible light waves that nobody expected were even there before Maxwell's equations predicted them. The rainbow-colored band is all we can see – everything else is invisible. Our eyes evolved to see this region with the peak sensitivity of our retina in the green – which is why green laser pointers look brighter than blue or red laser pointers of the same power.[78]

short-wavelength X-rays.[79] We now know there is a massive band of light waves outside the visible range, from radio waves to gamma rays – and this band or spectrum is called Maxwell's Rainbow (Figure 1.18). With the blood, sweat, and tears, of scientists working for over 150 years, we can at last conclude that Huygens was right, and Newton was wrong. Light is a wave, not a particle. Just kidding! In the next chapter, we will provide equally ponderous evidence that light *is* a particle and *not* a wave – at least some of the time. This rediscovered particle nature of light will be critical for its use as the carrier of information on the quantum internet. Not to worry, though – all the information in this chapter will be vital for us to understand the classical internet that we'll also discuss in future chapters.

78 A figure of Maxwell's Rainbow from https://commons.wikimedia.org/wiki/File:EM_Spectrum_Properties_reflected.svg, "A diagram of the electromagnetic spectrum," by InductiveLoad. This file is licensed under the Creative Commons Attribution-Share Alike 3.0 Unported license.

79 For more on Röntgen and his invisible *x*-ray waves, see https://en.wikipedia.org/wiki/Wilhelm_R%C3%B6ntgen. (He also won a Noble prize for his discovery.)

CHAPTER 2

What Light Through Yonder Window – *Breaks?*

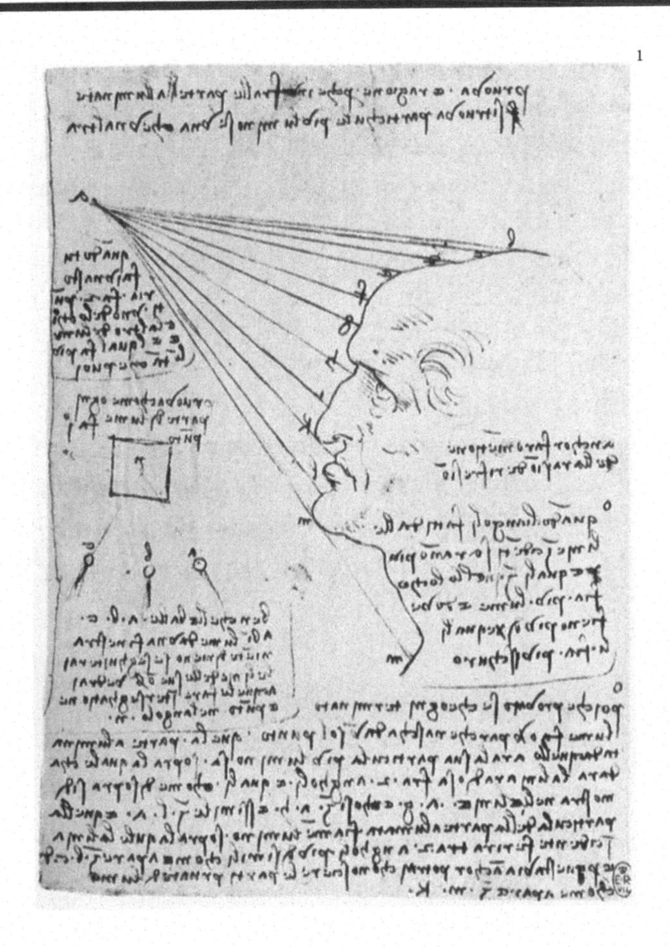
[1]

1 "Study of the effect of light on a profile head," by Leonardo Da Vinci. Work is in the public domain and is here https://commons.wikimedia.org/wiki/File:Leonardo_da_vinci,_Study_of_the_effect_of_light_on_a_profile_head_(facsimile).jpg.

I can't define a photon, but I know one when I see one.

Roy J. Glauber[2]

THE END OF PHYSICS?

By 1900, some scientists were predicting that it was the end of physics. We had Newton's Law of Gravity, Newton's Laws of Motion, and Maxwell's Equations, and all physical phenomena were, in principle, described by these laws and equations. Some claimed that the age of discovery was over, and physics was demoted to engineering – just putting all these equations to work. Who claimed this? For years, I had read that it was the British physicist, William Thomson (more commonly known as Lord Kelvin), who had become famous for his work in thermodynamics.[3] The quote that was bandied about was, "There is nothing new to be discovered in physics now. All that remains is more and more precise measurement." It turns out this is a misattribution. It is instead a paraphrased comment by the American physicist, Albert Michelson, who is famous for his experiments attempting to measure changes of the speed of light as it moved through the luminiferous aether.[4] In 1894, Michelson stated that:

> It is never safe to affirm that the future of physical science has no marvels in store which may be even more astonishing than those of the past; It seems probable that most of the grand underlying principles have been firmly established and that further advances are to be sought chiefly in the rigorous application of these

---

2 Roy J. Glauber is an American Physicist and a 2005 Nobel Laureate. He said this at a Workshop in Celebration of the 50th Anniversary of the Discovery of the Lamb Shift, 18–22 August 1996, Bellingham, Washington. I was there and took notes. It was in the question and answer session after the talk of another Nobel Laureate, Willis Lamb (the guy who first measured the Lamb shift), who was arguing that nobody, but he knew how to define the notion of a photon properly. Glauber's response to Lamb was a joke based on a quote by U.S. Supreme Court Justice Potter Stewart, who, in 1964, while trying a case on free speech vs. pornography, stated in court, "I can't define pornography, but I know it when I see it." See http://answers.google.com/answers/thread view/id/226615.html. The older members of the audience laughed.

3 For more on Lord Kelvin, and predictions he did get wrong, see https://en.wikipedia.org/wiki/William_Thomson,_1st_Baron_Kelvin. Of particular note is his aversion to human flight with his 1902 quote, "No balloon and no aeroplane will ever be practically successful." He also predicted that the Earth only had about 400 years of oxygen left and that the Earth was just 100 million years old instead of the commonly accepted 4.5 *billion* years old – but what's a few orders of magnitude between friends?

4 For more on Michelson, see https://en.wikipedia.org/wiki/Albert_A.Michelson.

> principles to all the phenomena which come under our notice ... An eminent physicist remarked that the future truths of physical science are to be looked for in the sixth place of decimals.[5]

Indeed, it would be strange for Lord Kelvin to have said anything like this. Around the same time, he gave a lecture on the trouble with physics and pointed out that things were not as nicely wrapped up as Michelson thought. In 1900, Lord Kelvin gave a speech entitled, "Nineteenth-Century Clouds over the Dynamical Theory of Heat and Light." The talk began:

> The beauty and clearness of the dynamical theory, which asserts heat and light to be modes of motion, is at present obscured by two clouds. The first came into existence with the undulatory theory of light, and was dealt with by Fresnel and Dr. Thomas Young; it involved the question, how could the earth move through an elastic solid, such as essentially is the luminiferous aether? The second is the Maxwell-Boltzmann doctrine regarding the partition of energy.[6]

It was clear from the speech that Kelvin thought the existing theories would eventually explain these two clouds. Little did he know that the clouds would turn into typhoons. Out of the first typhoon would speed Einstein's theory of relativity, and out of the second would leap quantum physics. We are not so interested here in the first cloud, but I will make a note that cloud had formed due to the experiments of Michelson, who I just quoted as saying, above, that no further physical

---

5 Michelson, A. A. "Some of the Objects and Methods of Physical Science." *The University of Chicago Quarterly Calendar* III (May 1894): 15. This quote was maddeningly difficult to track down, as many books misquote it, give the wrong reference, or give the right reference with the wrong date. Thank heavens for the Internet and Google books! The original is here https://books.google.com/books?id=B23OAAAAMAAJ.

6 Hon, Right. "Nineteenth-Century Clouds over the Dynamical Theory of Heat and Light." In *Notices of the Proceedings at the Meetings of the Members of the Royal Institution of Great Britain with Abstracts of the Discourses*, edited by Lord Kelvin, G.C.V.O. D.C.L. LL.D. F.R.S. M.R.I., vol. 16, 363–397. Royal Institution of Great Britain, January 1, 1902, W. Nicol, Printer to the Royal Institution. He gave the speech on Friday 27 April 1900, but it took two years to make it into print. Thanks again to Google books, you may find the original here: https://play.google.com/store/books/details?id=YvoAAAAAYAAJ. When Google declared some years ago that they planned to scan and make publicly available every book ever written, I frankly thought this would be an impossible task.

truths would emerge in the physical sciences. The old theory of light predicted that, since light was a wave and waves are always wiggles moving through something, there had to be some invisible and intangible stuff – filling all the universe – for the light to travel through. That stuff was called the "light-bearing intangible substance" or "luminiferous aether" for short. The Michelson–Morley experiment provided evidence that the aether did not exist, and Einstein proved that there was no need for it in the first place. Following that trail, Einstein then produced the theory of relativity. But that is a tale for another time.[7] For our purposes here, we need to understand what was the *second* cloud hanging over physics – What was wrong with the Maxwell–Boltzmann theory of the partition of energy? What the heck is the Maxwell–Boltzmann doctrine on the partition of energy, you might ask? Let's begin at the very beginning – the British Navy needed bigger and better steam engines.

The invention of the steam engine began the industrial revolution. Instead of using water mills, windmills, or animal muscles, you could do work by converting coal into heat to produce steam to drive an engine. Since coal is abundant and portable, the steam engine revolutionized shipping and the railroad industry.[8] Even an early automobile ran on steam power, and it was the steam power that ran the Titanic into an iceberg. The design of the steam engine rapidly settled down into the form shown in Figure 2.1. A coal fire heats water into steam, which is allowed to expand in a primary cylinder and push up on a piston, a movement called the power stroke. After the power stroke, the vapor cools and vents into a second cylinder that is in cold water. The steam compresses as it cools, creating a vacuum in the primary cylinder, and then the primary cylinder moves back down the other way in the induction stroke. You flip a valve to allow more steam in, and the process repeats.

It was the Scottish inventor, James Watt, who realized that the entire process worked by moving heat from a hot source (the coal fire) to a cold source (the secondary tank).[9] This steam-engine process is the essence of thermodynamics – by moving heat energy from a hot source to a cold sink, you could do work such as turn a ship's propeller. The quest was on for evermore efficient steam engines. The question was,

---

7 For more on the Michelson-Morley experiment and connections to Einstein's relativity, see https://simple.wikipedia.org/wiki/Michelson%E2%80%93Morley_experiment.

8 For the history of the steam engine, see https://en.wikipedia.org/wiki/History_of_the_steam_engine.

9 The biography of James Watt is here https://en.wikipedia.org/wiki/James_Watt.

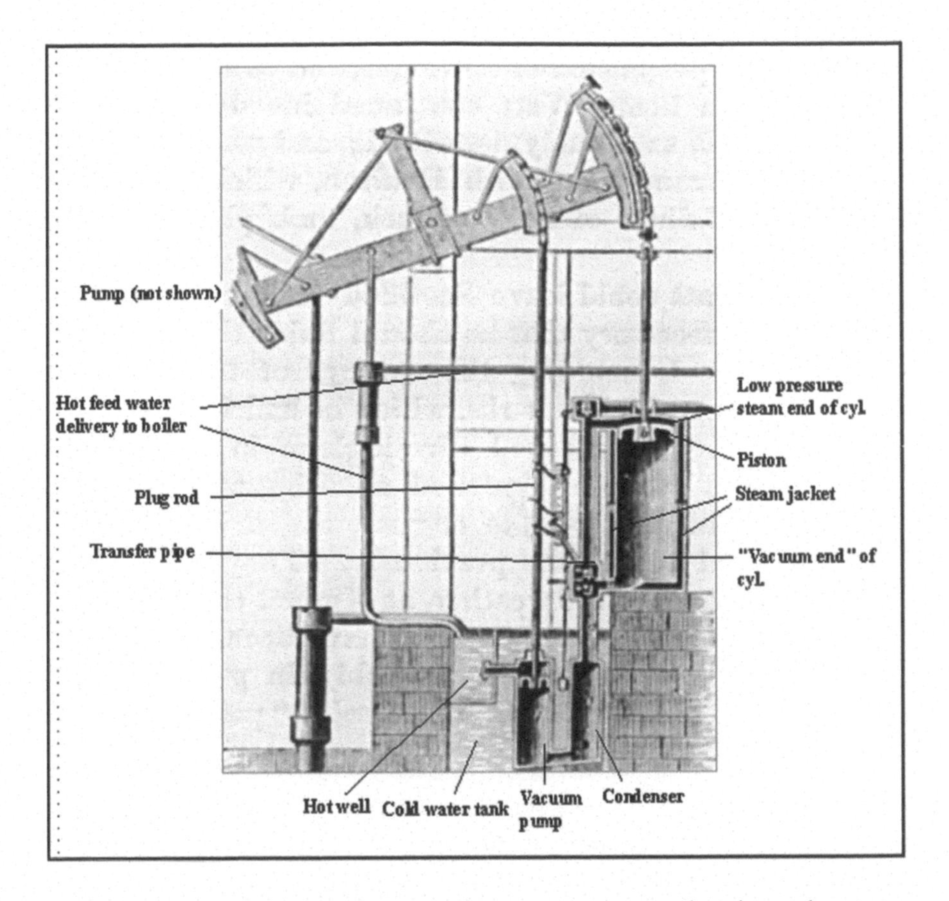

FIGURE 2.1 A schematic of an original Watt steam engine that shows the piston and hot source and cold sink. Watt constructed his first working model in 1765.[10]

given a certain amount of coal that, when burned, produced a certain amount of energy, how much of that energy could we use to do work? The first law of thermodynamics tells us that energy is conserved, and so you might think that is all there is to it. But there is the second law of thermodynamics that places another constraint – heat energy can only move from a hot place to a cold place but can never do the reverse. The reverse would be consistent with the first law but not the second.

10 A schematic of Watt's steam engine from *History of the Growth of the Steam engine.* By Robert H. Thurston and published by D. Appleton & Co., 1878. This media file is in the public domain in the United States. See https://commons.wikimedia.org/wiki/File:Watt_steam_pumping_engine.JPG.

These two laws came about in the study of steam engines, more generally called heat engines, where pistons and valves move heat from a hot place to a cold place. The second law tells us that not all of the energy stored in the coal produces work and that we must always waste some. One form of the second law is precisely that – it is impossible to build a perfectly efficient heat engine. The efficiency is a function of the temperature of the heat source and that of the cold sink.

If you have ever taken a course on thermodynamics, it is a mind-muddling morass of math that relates energy, heat, temperature, and pressure of such engines, without any real understanding as to what is going on. What *are* energy, heat, temperature, and pressure? Thermodynamics is a somewhat qualitative macroscopic theory of such engines based on many years of observations of their performance. We were missing an atomic-level approach. Missing until Maxwell once again came to our rescue. Of course, his bipolar colleague Ludwig Boltzmann helped.[11] In the mid-1800s, physicists were divided over the nature of heat and energy. Some viewed it as sort of a continuous fluid that flowed from place to place. But others, such as Maxwell and Boltzmann, took the point of view that heat was due to the motion of unseen atoms and molecules. We must note that even by 1900, a significant proportion of physicists doubted that atoms existed. Only in 1905, did Einstein finally lay the matter to rest, in favor of atoms, upon publishing his work on the chaotic motion of dust or pollen grains observed under a microscope.[12] So in the late 1800s, taking up the atomic theory was asking for trouble. Maxwell mostly kept his head down, but Boltzmann publicly debated the matter with those who thought that heat was a continuous and moveable fluid. The debates were rancorous, and some say it was this hatred that led Boltzmann to suicide. The atomic theory eventually won, and the continuous fluid theory of heat lost, and so we focus on the winner.

In the Maxwell–Boltzmann theory, the flow of heat was fundamentally the movement of heated atoms or molecules. The model assumes

---

11 For a biography of Boltzmann, see https://en.wikipedia.org/wiki/Ludwig_Boltzmann. Boltzmann most certainly had what we call bipolar disorder, and he ended his own life in 1906 by hanging himself while on vacation in Trieste: http://www.washingtonpost.com/wp-srv/style/longterm/books/chap1/lisemeitner.htm.

12 For more on Brownian motion and Einstein's miracle year, see https://en.wikipedia.org/wiki/Brownian_motion.

that you have atoms, something like ping-pong balls, rattling around in that steam-engine piston. When heated, the atoms convert the heat into kinetic energy and move around more frantically as they get hotter – and that kinetic energy does work to move the piston as the atoms collide with it. In the other direction, they move more slowly as they cool down, and as the temperature approaches absolute zero, they nearly stop. (The third law of thermodynamics states that you can never actually reach absolute zero, but we have achieved, in the lab, temperatures of one-millionth of a degree above absolute zero.) The atom theory accounts for the first law, that energy is conserved. Heat energy is transferred into the kinetic energy of atom motion and back again.

It is a bit trickier to extract the second law from the atom point of view and attempts to do so drove Boltzmann mad. At the atomic scale, the equations of motion for ping-pong balls are invariant under time reversal. That is, the equations do not contain an arrow of time. Run a film of atoms moving forward in time, and it looks just as physical as a film of them running backward in time. But the second law contains an arrow of time – heat always flows from hot to cold but never the reverse. If you place a cup of boiling tea on your desk and wait a while, eventually, it becomes the same tepid temperature as your room. Never do you observe that when you place a cup of lukewarm tea on your desk, the tea heats up and starts to boil. That is the second law in a nutshell. Physicists still argue over this arrow of time business and write books about it, but we shall not take that detour here – for that is the road to perdition.[13]

Mathematically, one of the most natural systems of heated atoms to study is a system in thermal equilibrium. In our example with the teacup, the boiling hot tea cools and eventually becomes the same temperature as the air in the room. At that point, the tea neither heats nor cools – it is in thermal equilibrium with the air. If the room is at 26°C, so will be the tea. There is no net heat flow on average. The molecules in the tea and the molecules in the air are still moving around like crazy and colliding with each other, but no net heat transfer occurs from the air to the tea or vice versa. To simplify things further, Maxwell and Boltzmann replaced the tea with an enclosed cylinder of

13 See Lineweaver, Charles H., Paul C. W. Davies, and Michael Ruse. *Complexity and the Arrow of Time*. Cambridge University Press, 2013. which is here http://www.worldcat.org/oclc/934395585.

an ideal gas – treating the atoms like ping-pong balls that exchange heat energy by colliding with each other, but with no net heat traded on average. For a fixed temperature, some of the atoms will be moving a bit faster than average, and some atoms will be running a bit slower, but the average speed will be a constant that does not change. We call the distribution of speeds around this average the Maxwell–Boltzmann distribution, as mentioned in Michelson's commentary above. Because the system is so simple, there is an exact formula for the distribution.[14]

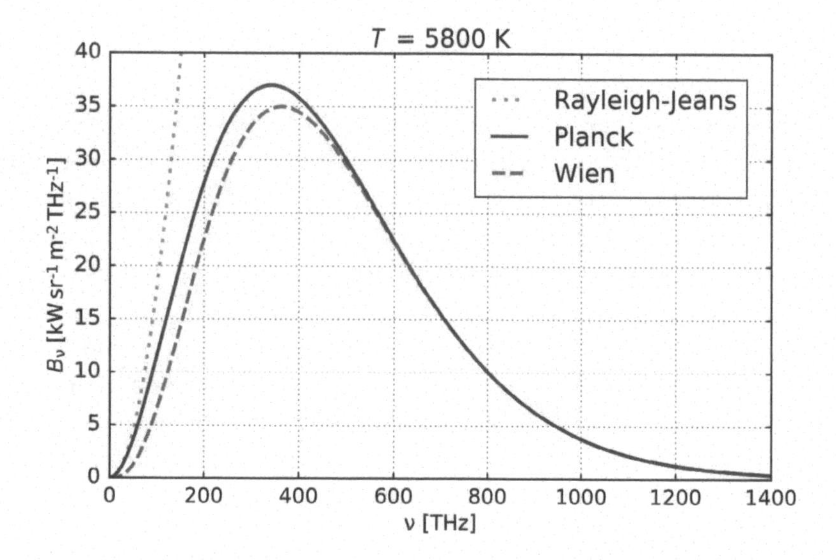

FIGURE 2.2 The three theories of the blackbody radiation. The plot shows the emitted light power coming out of a hole poked in the side of the blackbody. The plot is in frequency, so ultraviolet light is to the right at high frequencies, and infrared light is to the left at low frequencies. The dotted curve is the Rayleigh–Jeans theory, which goes to infinity at high frequencies. The solid curve is the Planck theory that agrees spot-on with experiment. The dashed line is the Wien approximation that agrees well at high frequencies but poorly at low frequencies.[15]

---

14 For more on the Maxwell–Boltzmann distribution, see https://en.wikipedia.org/wiki/Maxwell%E2%80%93Boltzmann_distribution.

15 This file is licensed under the Creative Commons Attribution-Share Alike 4.0 International license and is here https://commons.wikimedia.org/wiki/File:Mplwp_blackbody_nu_planck-wien-rj_5800K.svg.

The theory makes a precise prediction that the average kinetic energy of the gas molecules is proportional to the temperature. This notion makes sense, as you heat the gas, the energy-of-motion increases. This result is called the equipartition theorem, which relates the temperature to the average kinetic energy. The formulas agree very well with experiment, and they only rely on Newton's equations of motion for the colliding atoms and a smattering of statistics. Since this theory worked so well for atoms, in the late 1800s, physicists decided to apply it to light waves in a box at thermal equilibrium. It was then that things went horribly wrong.

PLANCKING FOR BEGINNERS[16]

The analog of the cylinder with atoms is light waves in a box in thermal equilibrium with the walls. Again, the theory is quite simple for thermal stability – when the walls and the light waves are all the same temperature, and the amount of heat from the walls flowing into the light, and from the light back into the walls is, on average, constant. We treat the walls as made up of oscillating antennas (a good model for atoms) that can either emit light waves or absorb them. We hold the walls containing the antennas at a fixed temperature, and the light waves should also then be at the same set temperature as the walls. The experiments were quite easy to do too. You take a hollow chamber and coat the inside wall with charcoal, which is a good absorber, and hence a good emitter, of light waves of heat. Then you heat this thing, and the charcoal goes from black to red hot. The red-light waves are being emitted into the chamber and then being absorbed by other charcoal molecules in the walls. Since light waves don't bounce off each other, like ping-pong balls do, the theory is straightforward. The goal is to find a simple distribution of the energy of the light waves in thermal equilibrium and to compute the average power at a fixed temperature.

The British physicists, Lord Rayleigh (in 1900) and James Jeans (in 1905) computed this distribution of energy per wavelength, and the

16 I take everything in this section from the masterpiece, Kuhn, Thomas S. *Black-Body Theory and the Quantum Discontinuity, 1894–1912.* Oxford: Clarendon Press, 1978. http://www.worldcat.org/oclc/976855047. This fellow is the same Thomas Kuhn who wrote, The Structure of Scientific Revolutions, where he introduced the term "paradigm shift".

result was – *complete and utter nonsense!* The theory predicted a distribution for the light energy that blew up to infinity as the light waves got shorter and shorter. This explosion meant the total integrated power was infinite. Take this in for a second. The theory implies that if you take a pottery kiln, paint the inside black, poke a hole in the side, and then turn on the furnace, an infinite blast of ultraviolet light emerges from the tunnel to destroy the entire universe. This theoretical disaster was called the ultraviolet catastrophe.

Luckily for us, no such catastrophe is seen in the lab. As you turn the kiln on and the temperature rises, you observe, for a fixed temperature, there is an average energy for the emitted light at a particular wavelength, say red hot. The probability of the ultraviolet light coming out drops to zero exponentially so fast that the total energy is finite. Since we always cover the inside of these kilns with soot or charcoal or black paint, they are known as blackbodies, and the Rayleigh–Jeans blackbody theory failed big time. The measured distribution looked a lot light the Maxwell–Boltzmann distribution for ping-pong-ball atoms, a Gaussian or normal-distribution looking thing with a peak at the average energy. (See Figure 2.2) Unlike the Maxwell–Boltzmann law, the experimental distribution showed the maximum occurs at a point where the frequency of the light is proportional to the temperature. The frequency of light is inversely proportional to the wavelength, so this result implies that as the temperature goes up, the wavelength corresponding to the peak gets shorter, going from infrared to dark red to eventually orange and then on to yellow hot. However, the nonsensical theory had only two ingredients to it, Maxwell's theory of light and Newton's theory of particle motion. These two ingredients gave rise to nonsense. Either one or the other – or both – had to be wrong! Lord Kelvin's cloud was a raging hurricane battering on the very shores of the foundations of physics.

As can be seen in Figure 2.2, the Rayleigh–Jeans theory (dotted line) agrees quite well with the correct approach, and the data (solid line), at very long wavelengths or minimal frequencies. That is, it concurs at the red end of the spectrum but fails horribly at the ultraviolet end. In 1896, the German physicist, Wilhelm Wien (pronounced Vill-helm Veen) discovered a distribution qualitatively for the blackbody spectrum that did the reverse, it agreed with the data at short wavelengths but disagreed at long wavelengths where the Rayleigh–Jeans curve worked best. Called the Wien distribution, it was more of a lucky guess than a real theory, but it fit the data quite well (Figure 2.2).[17] It would turn out, as Einstein would show several years later, that the Wien theory was equivalent to assuming that the light waves were particle-like. Recall that we just spent an entire chapter providing evidence that light was *not* particle-like. What the heck is going on? We have two theories for the blackbody radiation – the rigorous Rayleigh–Jeans law that fits well at long wavelengths but gives nonsense at short wavelengths, and the kludged-up Wien law that fits well at short wavelengths but disagrees with the data at long wavelengths. It was then that German physicist, Max Planck, literally, had a fit.[18]

Planck was unhappy with both of these theories but noticed the excellent agreement of one approach at one end and the other method at the other end. When something like this happens, you can force them into a mathematical shotgun marriage by fitting them together. Typically, this involves a fitting parameter that is adjusted so that the newlywed theories agree with the data. Since the Rayleigh–Jeans and the Wien formulae were particularly simple and depended only on the temperature, wavelength, and a handful of scientific constants, Planck was able to mash them together using a single fitting parameter he called h. This constant was then adjusted to agree with the data, and agree with the data it did, and the Planck distribution was born. (See the solid curve in Figure 2.2.) He did the math, but what was the physics? Planck spent the next few years trying to figure that out. Every time he tried to derive his new formula from basic principles, he would always end up back at the absurd Rayleigh–Jeans result. It was then that Planck had to choose who was wrong – Newton or Maxwell? Planck picked Newton.

The only way out of the mess, and the only way to recover his blessed and beautiful data-fitting distribution, was to make a startling claim. The little atomic antennas that made up the walls of the blackbody could not emit light at arbitrary frequencies but rather only at discrete frequencies. This assertion is like saying that your car radio can only receive radio stations at 89 or 90 megahertz but never at 89.3 megahertz. Since 89.3 megahertz is my favorite local public-radio station, that would be very sad. Since every antenna is as good as emitter as it is

17 For more on Wien and his distribution, see https://en.wikipedia.org/wiki/Wilhelm_Wien.

18 For more on Professor Planck, see https://en.wikipedia.org/wiki/Max_Planck.

a receiver, Planck's postulate would also mean that the broadcast tower, too, could never broadcast at 89.3 but only at 88, 89, 90, and so forth, in only whole numbers of megahertz. For classical antennas, that notion was nonsense. But something had to give to get the blackbody curve right, so Planck postulated that for tiny, atomic antennas you got 88, 89, 90, and so forth, but never anything in between. These integer frequencies were Maxwellian heresy. Planck called these integer chunks of energy quanta (quantum for the singular).[19] And, thus was quantum theory born. Planck did try desperately to save Maxwell's theory and proclaimed that the light waves were still able to take on any frequencies but that only the atoms were weirdly transmitting and receiving energy in chunks. Einstein then proceeded to kill off Maxwell too.

In 1905, Einstein published three papers in the same volume of the German journal, the *Annals of Physics.* That year is called Einstein's miracle year. Any *one* of those three papers could have won him the Nobel Prize, but only one of them did, his article on the blackbody theory. (This particular volume of *Annals of Physics* is a collector's item and one of the most stolen books from any library.)[20] The paper we have in mind here is his photoelectric effect paper. Still, the photoelectric effect theory (for which Einstein won the Nobel Prize) was a brief note at the end of an article titled, "On a Heuristic Viewpoint Concerning the Production and Transformation of Light."

Einstein began with a detailed analysis of the Planck distribution. Instead of the atoms in the walls, he focused on what Planck's theory implied for the electromagnetic light field inside the cavity. He performed a statistical analysis of the fluctuations of the light field and made an astonishing discovery. At small frequencies, in the Rayleigh–Jeans regime, the fluctuations had the form one would expect from waves. But at high frequencies, in the Wien regime, the fluctuations were more characteristic of what you would expect for particles. Einstein was able to derive the well-fitting Wien approximation by considering the light field to be made up of only classical-like noninteracting particles.[21] From this analysis, he made a daring leap and claimed

---

19 The word quantum comes from the Latin word *quantus*, meaning "how great" and it is the same root for the English word "quantity." A quantum leap is very small and not very big at all.

20 For more on Einstein's miracle year, see https://en.wikipedia.org/wiki/Annus_Mirabilis_papers.

21 Some years later, in 1924, the famous Indian physicist, Satyendra Nath Bose, was able to treat the light field as consisting of a particular type of non-interacting quantum particle. He showed that

that the light field, at least in the high-frequency regime, consisted of *particles!* Einstein resurrected Newton's particle theory of light from the dead. However, there was an added twist – the light particles or quanta of light had energy proportional to their frequency, and the proportionality constant was Planck's constant h.[22] Maxwell's theory had nothing like this. The energy of a classical wave is proportional to the square of the amplitude or the intensity of the wave and not related to the wave's frequency or wavelength. Down went Maxwell's theory of light. Then finally, in a one-off, Einstein concluded his paper by observing his light quanta hypothesis could explain the photoelectric effect. For that, he won the Noble Prize.

In 1902, Einstein's archenemy, the German physicist Philipp Lenard (an anti-Semite and soon-to-be Nazi), carried out a series of experiments. He showed that light kicked an electron out of an atom in a way that the energy increased with increasing frequency of the light. It did not increase with the growing light intensity, which is what Maxwell's theory predicted. Scientists christened this oddball behavior the photoelectric effect, and it is the physics behind digital cameras and laser copying machines. It was as if you were playing a game of pool, and the light field was the cue ball knocking the eight ball off the pool table. That doesn't sound much like a wave. Einstein's 1905 theory of light quanta fit Lenard's data right on the nose, and it was with this experimental agreement that Einstein ended his photoelectric-effect paper.[23] Lenard the Nazi did not like that much, since Einstein was Jewish, and he did his best to block Einstein from winning the Nobel Prize – for any discovery at all! Lenard and his cronies succeeded for a while until, by 1921, Einstein was so famous that it was embarrassing that he did not have a Noble prize yet. The Swedes finally came around and gave it to him for his theory of Lennard's photoelectric effect experiment. Einstein went on to become world-famous, and invading allied forces expelled Lenard from his university post in 1945.[24]

---

you could get exactly the Planck distribution and not the approximation. See https://en.wikipedia.org/wiki/Bose%E2%80%93Einstein_condensate#History.

22 People often refer to □ = $h/(2\pi)$ also as Planck's constant, but this is called Dirac's constant. Dirac got his constant by dividing Planck's constant by 2π.

23 For more on the photoelectric effect, see https://en.wikipedia.org/wiki/Photoelectric_effect.

24 Walter Isaacson juicily describes this story in his splendid biography, *Einstein – His Life and Universe*. New York: Simon & Schuster, 2007. http://www.worldcat.org/oclc/76961150. Einstein died on April 18th, 1955 – the day after they baptized *me* – and his estate sealed his private papers

These light-energy quanta have a more common name – they are called photons.[25] We must reconcile Einstein's idea with the Newton–Huygens wave-particle wars of the previous two and a half centuries. Newton and Huygens argued that light must be either a wave *or* a particle – that was their mistake. The quantum resolution is that light is *neither* a wave *nor* a particle. The idea is that some experiments reveal the wave nature of light, and others reveal the particle theory of light. Until the blackbody and photoelectric-effect experiments, there were no tests that probed this particle nature, and so the wave theory had fit nicely. Also, most classical light sources, say a laser pointer, consist of billions of photons, so the behavior of the individual photon is hidden.

But how can something be neither a wave nor a particle until it is measured? It is like saying that sometimes when you look at a ripple on a pond, it seems like a ripple on a pond, but other times when you look, it looks like a cue ball on a pool table. What does that even mean? It is at this juncture that most authors on this topic would drag you back to the Young two-slit experiment and try to make sense of it there. But I have grown weary of this explanation, and I surely don't want to Bohr you, and so I will give you a series of more modern experiments. In each case, I will present the classical wave theory and an experimental result (good for laser pointers) and then the quantum theory and experiment (suitable for single photons). These are not thought experiments, but actual tests carried out since the 1960s. The classical theory will make intuitive sense, but the quantum theory will be bizarre. You probably won't like the quantum theory much, but it's not my job to make you like it. It is my job to make you *understand* it. The quantum theory and associated experiments are the reality – and reality doesn't care a whit if you like it or not.

ONE PHOTON, INDIVISIBLE, WITH QUANTA OF ENERGY, IN BALLS

When I began my Ph.D. research, my advisor suggested I look at the theory of one photon in empty space. I rejected this problem as too

---

for 50 years. Isaacson was the first to access those papers, in 2005, and then to compose this biography with all the details hidden for so long.

25 The origin of the name photon can be found here https://en.wikipedia.org/wiki/Photon.

complicated and spent four years studying the properties of empty space itself before I would work up enough courage to add a single photon. In my defense, the quantum vacuum of empty space is not empty at all, is very complicated, and appears to have infinite energy.[26] The quantum theory implies that photons are essential objects that we cannot partition. That is, you can have one or two or three photons but never half of a photon. The discreteness of a photon is a theoretical prediction of the quantum theory, and we must test it. We must also compare it to the prediction of the classical theory, which holds when the beam of light contains billions of photons, such as from a laser pointer. We will first start with the classical theory and experiment shown in Figure 2.3.

In Figure 2.3, a laser of intensity one (in units of one milliwatt per square centimeter) shines on a half-silvered mirror (a 50–50 beam splitter) that splits the beam precisely in half. The two detectors each record an intensity of one-half. This experiment can be explained

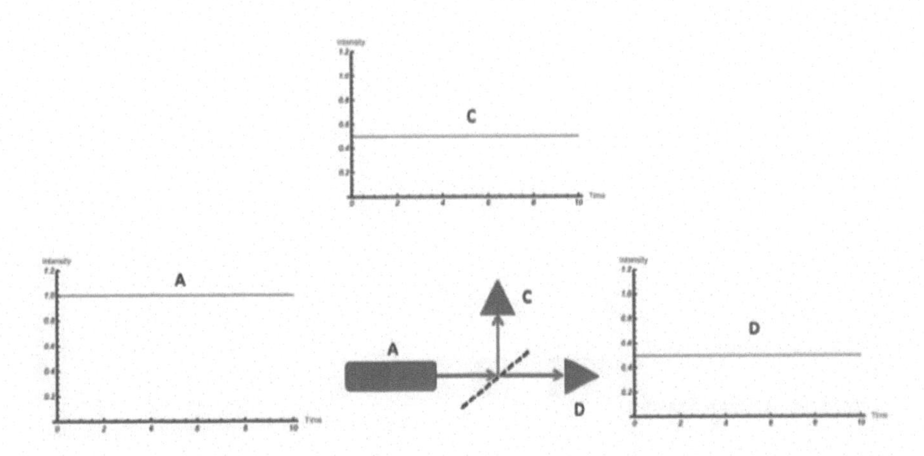

FIGURE 2.3 A rectangular-shaped laser (A) (classical light source) shines light on a half-silvered piece of glass (called a 50–50 beam splitter, dashed line). The light splits in half and then registers on the two triangle-shaped detectors C and D. The laser emits light with intensity one (in some units), as shown on the left graph A. The two indicators each register a power of a half, as shown on the charts C and D.

26 Dowling, Jonathan P. "The Mathematics of the Casimir Effect." *Mathematics Magazine* 62, no. 5 (Dec. 1989): 324–331. See http://www.jstor.org/stable/2689486.

equally well with either the Newton particle theory or the Huygens–Fresnel wave theory. In the wave theory, the intensity is related to the height of the waves. In the particle theory, we associate the power with the number of particles flowing out of the laser per second. For a laser pointer, this can be a billion particles per second. In either theory, this experiment makes perfect sense – half the light energy goes one way, and half goes the other way. To compare, in Figure 2.4, we show what happens when we replace the laser with a single-photon gun that fires only single photons at equally spaced time intervals. Such photon guns exist in the laboratory, and the experimental results agree with the quantum theory of light. Something different happens in this experiment. The wave theory cannot explain the analysis at all. The light field consists of single photons, and no wave theory is needed. Indeed, the wave theory would predict that the single photon splits in half, with one half going one way, and the other half going the other way – but this

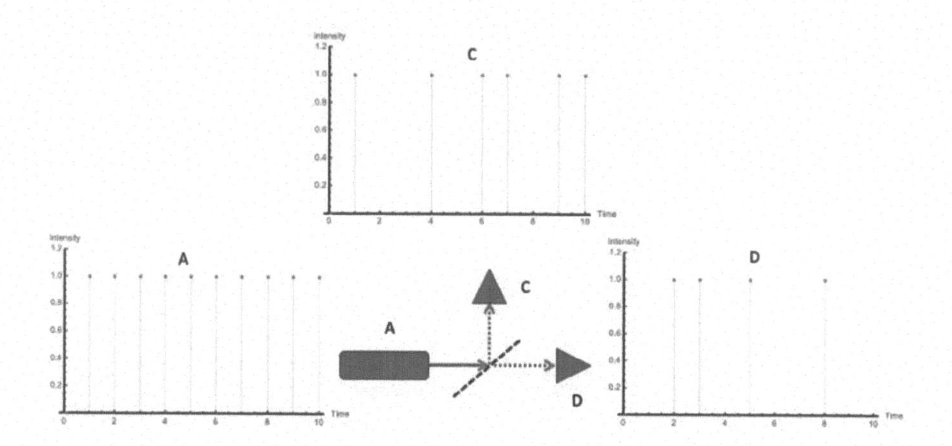

FIGURE 2.4 Here we show the particle version of the experiment in Figure 2.3. Instead of a laser pointer on the left, we have a single-photon gun that shoots out a single photon each nanosecond (A). The photon is indivisible. Hence when a photon hits the 50–50 beam splitter, the upper detector has a 50% chance of going click, and the lower detector has a 50% chance of going click. But if one detector clicks, the other cannot click. The Newton particle theory states that the photon either goes up or to the right. The quantum photon theory says that the photon is a superposition of neither going up and nor going to the right, like Schrödinger's cat, until a click occurs, and the cat state collapses, with upper click "dead" and lower click "alive".

experiment rules out ever seeing half a photon. The entire photon all goes one way, or it all goes the other way. A photon is an indivisible object, like the indivisible atoms of Leucippus and Democritus.

If the photon was a Newtonian particle, when it hits the beam splitter, it can either go straight through or bounce off, with a 50–50 probability. The single-photon gun fires the photons once per second: bang-bang-bang-bang-bang-bang-bang-bang-bang-bang, for the ten photons shown in Figure 2.4. In this particular run of the experiment, the upper detector C goes click-blank-blank-click-blank-click-click-blank-click-click, while the lower detector D goes blank-click-click-blank-click-blank-blank-click-blank-blank. That is, whenever C goes click, D does not click, and vice versa. If we were to run the experiment over again, we'd get an entirely different set of clicks and blanks, but each time the click in one detector would correspond to the blank in the other sensor. If we were to run the experiment a billion times, we would recover the classical result in Figure 2.3, since the average of one (getting a click) and zero (getting a blank) is one half, which is the intensity seen at the detectors in Figure 2.3. This second experiment demonstrates the indivisibility of the photon, which is only explainable in the particle theory. The wave theory would predict that you would get half of a photon at each detector. We never see that, but if you average over billions of runs, then you see half the intensity at one indicator and half the power at the other. Only in the limit of vast numbers of photons does the wave theory seem to work.

If the upper detector gets a click, the lower detector cannot get a click. The clicks on the lower detector correspond to when the top indicator gets no clicks. That is the result predicted by the Newton classical particle theory and the quantum photon theory. However, the *interpretation* is radically different between the two approaches. In the Newton particle theory, if you could know exactly the initial position and velocity and angle of the light particle as it leaves the gun, and if you knew all the details of the roughness and construction of the half-silvered mirror, you could predict with certainty where each light particle would go – up *or* to the right. However, since it is difficult to know all these things in practice, we settle for a classical statistical theory – on average, half the light particles go up, *or* half go to the right. It is too complicated to compute on an individual basis what particle will go where.

Compare this to the quantum theory interpretation – in the quantum theory, the photon goes *neither* up *nor* to the right. Like Schrödinger's cat, which is neither dead nor alive, a single photon is neither going to the right (dead) nor going up (alive).[27] Only when one of the detectors clicks does the photon collapse into a particle and decide which way it went. If the upper detector clicks, then the lower one cannot click, and the photon is "dead". If the right detector clicks, then the upper detector cannot click, and the photon is "alive". These are the quantum postulates of unreality and uncertainty at work. After the beam splitter, the direction the photon goes is *unreal* – it has no path until it is measured. The direction the photon goes is also *uncertain* – *which* detector clicks cannot be predicted, not even in principle, no matter how much information you have about the experiment, the gun, the beam splitter, and so on.

This quantum uncertainty makes for an excellent and commercially available quantum random number generator. Real random numbers are hard to make in practice. A truly random string of zeros and ones can often be tough to make. Genuinely random numbers have applications to cryptography, computing, and communications technologies. The quantum fix is to take the experiment in Figure 2.4 and bin the detections as follows – if the upper detector clicks call that a "one-bit". If the lower detector clicks call that a "zero-bit". As you run the experiment, it produces a time sequence of genuinely random zeros and ones. Heisenberg's ghost and his uncertainty principle guarantee the randomness. Not just theory, these quantum random number generators are big business, and many companies are making them and selling them. Here is a simple example of photonic quantum technology at work.

IT'S A WAVE! IT'S A PARTICLE! IT'S – WHAT THE HECK *IS* IT!?[28]

While I claim that the quantum interpretation of the experiment in Figure 2.4 requires the assumption that the photon goes neither way,

27 Most often, you'll hear that the cat is *both* dead *and* alive. That is not accurate, since the cat has no deadness or aliveness until we make a measurement. Before the measurement, a better statement is to say it is *neither* dead *nor* alive. It could also be neither half dead nor half alive. This is quantum unreality. The cat does not have a definite state until we make the measurement. The cat before the measurement is sort of like a zombie – the living dead!

28 In this book, I have tried to avoid the notion of wave-particle duality to explain the double-slit or any other experiment. That is because I find wave-particle duality an archaic holdover from the early years of quantum theory. If you must know what it is, please read this, https://en.wikipedia.

org/wiki/Wave%E2%80%93particle_duality. More fun it to read this debate on the topic between my quantum physicist friends and me.

Chris: "Quantum physics friends: is wave-particle duality useful? I mean, quantum physics is basis of all technology. But one is not always simply solving the Schrodinger equation. We make use of concepts, analogies, approximations, etc. within the theory to make progress (even if we go back to the Schrodinger equation to work out the details). For example, I gather that a lot of material physics can be understood and progressed with simple pictures of energy level diagrams. Of course, the details then need the full theory – but the ideas are framed only by discrete energy levels. So, the discreteness of quantum physics is a useful abstract concept. But what use is wave-particle duality? Does it (abstracted from the rest of theory) provide a conceptual basis for anything practical, current or proposed?"

Richard: "Hmm, I'm very rusty now, but in the old quantum chemistry days I looked at symmetric tunneling splitting in molecular spectroscopy (e.g. ammonia): Wave-interpretation → Tunneling between seemingly identical ground states → interference in ground state → Observable splitting of the ground state energy (spectroscopy). Particle-interpretation → Tunneling probability becomes smaller as mass increases → heavier molecules have smaller splitting around ground state. Probably not what you were asking for. But 'both' effects are observable. Physical chemists use this argument to explain splittings (and their size) in molecular spectroscopy."

Howard: "Yes. Optical or matter-wave Interferometers with independent particles and click detectors."

Chris: "I'll accept this 😀. Follow-up: are interferometers (with click detectors!) used in any technology that is not used in state-of-art physics labs? Proposed technology includes, e.g., sensors?"

Aephraim: "Howard I buy tunneling etc., though more as a wave effect than as 'duality' per se; but interferometers with 'click' detectors I also see as just a wave effect. When particle counting is used, it's just because one wants more sensitivity (which is only bounded because of 'duality'). I'd kind of say once we navigate with atom interferometers, I'd consider that closer. Or use pair interference, e.g. HOM, maybe for those medical imaging applications?"

Rob: "I find the concept distasteful, as it is then too easy to slip into the language of a psi-ontologist."

Mankei: "Most photonics textbooks describe absorption, spontaneous emission, and stimulated emission using only Einstein's photon picture and rate equations. It's sufficient for all practical purposes in modeling lasers, light-emitting diodes, photodetectors, solar cells, etc. The behavior of light in resonators, lenses, waveguides, etc. is still modeled using classical EM, i.e., wave optics. Most people don't bother with the full quantum theory in electrical engineering. Also check out any undergraduate textbook on semiconductor devices; they are full of heuristic discussions of electrons/holes using the wave or particle picture depending on the phenomenon."

Howard: "Given that quantized energy levels is explained as a wave phenomenon, arguably WPD underlies everything useful."

Chris: "I don't understand where the duality comes in here? Sounds like everything as a wave works?"

Howard: "Chris when you ionize an atom you get a click. And there is all the other evidence that electrons are particles."

Jonathan P. Dowling: "WPD is a bunch of crap. I never even teach it in class."

Daniel: "Jonathan P. Dowling indeed: it's a heuristic/empirical argument used to make second quantization seem mysterious."

Jonathan P. Dowling: "Daniel, at last we agree on something!"

Daniel: 😲

Jonathan P. Dowling: "It is a relic of physicists struggling with how to fit archaic notions of 'wave' and 'particle' into the new quantum paradigm. You can do quantum physics without ever resorting to either notion."

Graeme: "I think wave-particle duality is the best way to understand quantum error correction."

Jens: "I do not think that picture is actually that useful. It is a historical picture that is in some ways misleading and confusing: Why and how should a quantum system decide whether it shows wave-like or particle-like features?"

Aephraim: "'Decide,' of course not. To me, 'wave-particle duality' doesn't mean that, but merely that the things we call electrons (or photons or …) share some properties classically associated with waves and some of those associated with particles – principally, that while their equations of motion have the form of wave equations (with tunneling, interference, and all that), they can't be absorbed ('fully detected') in two places at once. Not to make second quantization 'sound mysterious,' but merely to sum it up in words. Maybe 'field-particle duality' would be a more modern phrasing. The world is fields, but the occupation of the fields is quantized. Basta. And many above pointed out, as though it would be new to Chris, cases where we clearly see that these fields are quantized. But I think that does miss the question. How many of these applications really require the particle nature? I guess one can equivalently ask: how many of these applications is second quantization indispensable for describing? As I said before, Hong-Ou-Mandel, and hence linear-optical logic gates and the like, as well as applications to metrology/lithography (cough cough), et cetera. I'm trying to think of good examples with massive particles, e.g. atoms. Part of me wants to say anything beyond mean field would count. But thinking of applications where the particle nature matters … I'm having trouble. I guess some would say atom-interferometer gravimeters/accelerometers count, but that's stretching the duality to refer to any massive field, leaving out the actual quantization. Interaction-free measurement, of course ☺. Oh … à la IFM, anything else which requires discrete detections, so all of quantum communications. Of course, it doesn't usually require 'waves' (unless you extend your definition of 'wave' to include a field in any abstract space, i.e., polarizations) – but a Franson-based crypto system most definitely uses wave-particle duality inherently. PS Strike 'all,' perhaps, though we can argue about whether even CV crypto relies on discrete detections – just not discrete particle detections …"

Jonathan P. Dowling: "WPD completely breaks down when you are describing many-particle quantum entangled states."

Neil: "Is the Fourier transform useful?"

Aephraim: "Well, I've never seen anyone tax it …"

Neil: "On second thought, maybe I'll stay out of these debates and head back to the lab ☺."

Chris: "Don't forget your Fourier cheat sheet ☺."

Daniel: "It originated with Napoleon's attempt to conquer Egypt: trying to figure out how fast you can make a cavalry horse gallop without it dropping dead … (might have worked too, if it wasn't for that pesky Nelson guy …)"

Michael: "Not useful. Back when I used to tutor 1st year physics I'd ban reference to it!"

Jonathan P. Dowling: "'Physical reality is a ray in Hilbert space." – P. A. M. Dirac,'"

Chris: "Tell Ray dad says hi."

Jonathan P. Dowling: "I used to write a weekly column in the University of Colorado newspaper called, 'A Ray in Hilbert Space'."

Nathan: "Funny name for a newspaper!"

Aephraim: "That's what I used to call my thesis advisor. (Your column would've made more sense if you'd changed your name, taken one for the team.)"

Gavin: "One day bulk/boundary duality will be as second hand and as wave/particle duality – take your pick, use wavelet transform or a Fourier transform."

after the beam splitter, the demonstration does not reveal that assumption. It is entirely consistent with the Newton particle interpretation that the photon goes one way or the other. To expose this quantum

---

Bryden: "Like quantum therapy?"

Barak: "I'm pretty sure quantum field theory makes wave-particle duality obsolete."

Aephraim: "To me that's like saying QM makes Newton's third law 'obsolete'. Different concepts are useful in different situations and at different levels of description, and even if one is incomplete and derivable from another, that doesn't make it useless as one sort of description … I also don't harp a lot on 'duality,' but as part of a historical development of the theory & phenomena of QM, I don't think it needs to be erased either …"

Mankei: "I can now see why the engineering departments hate physicists teaching their courses. People who think wave-particle duality is not useful should try to explain something like https://en.wikipedia.org/wiki/Drift_velocity to undergraduate students. All your cute spherical-cow exercises with the Schrödinger equation won't get you anywhere close to the right model. You'd have to teach open quantum systems/quantum Langevin equations first if you want to stick to the full quantum theory. The only way that won't cause a student riot is to tell them it's okay to consider electrons as particles in a Brownian motion in this context. Of course, it is an approximation that fails under certain conditions and good scientists should recognize that, but this sort of heuristic picture is useful even for practising physicists/engineers so that we don't get bogged down with unnecessary details."

Nathan: "The biggest thing is that we need to stop putting forward particle/wave duality as a deep mystery of quantum mechanics that cannot be truly understood."

Matthew: "Agreed. However, we also shouldn't do that to a point where the 'magic' and fascination of QM isn't emphasized either. I've been working with entanglement for over 10 years and I still don't really 'get it'. Getting my head around other concepts in QM can take many years of experience. I love it for that."

Aephraim Steinberg: "Yes. I certainly tell my intro quantum students what people have meant by 'wave-particle duality,' but go on to say I dislike the term, 'wave-particle unity' would be preferable, because we're learning that quantum objects follow a clear set of rules, and it just happened that historically we termed some 'wave-like' and others 'particle-like'. I even briefly mention 'delayed choice,' because it always amazes the students, and because I think this means it drives home facts about interference they don't appreciate earlier – and then I say that personally I think the discussion is based on the wrong way of thinking about things (as though light 'chooses' to be a wave or a particle!), but that Wheeler was a brilliant man, and that they can think about things and form their own opinions. Re drift velocity, Mankei, why is this duality? Why do we need to think of waves at all in that context?"

Mankei: "Because those electrons/holes are near the band edges when you deal with semiconductor devices, and you can't explain energy bands and bandgap without the wave picture. People draw electrons/holes as particles moving in energy band diagrams, like the ones here: https://en.wikipedia.org/wiki/Band_diagram."

Jonathan P. Dowling: "Take an H-polarized photon and detect it in the H-V basis and you always get an H. Measure it in the ±45° basis and you get +45° or −45° at random. How do I explain this experiment with wave-particle duality? I can't. I could make up H-V / ±45° duality which states that you can't measure the H-ness and the 45°-ness of a photon in a single experiment. Put that way, the thought experiment reveals W-P duality for what it really is – utterly incomprehensible gibberish. Replace W-P duality and then you have a unifying theory. You can't measure conjugate observables in the same experiment. That is a unifying principle. W-P is a one trick pony."

weirdness, and rule out the Newtonian theory, we must construct an experiment where we measure the wave and particle nature of the photon in the same test. The key ingredient that made the Fresnel wave theory different than the Huygens theory was wave interference. Sometimes the waves add up, as in constructive interference, and sometimes they completely cancel out, as in destructive interference. One device that reveals this property is called an interferometer. A workhorse for the quantum theory of light is a particular device called a balanced Mach–Zehnder interferometer, named after the two physicists who thought it up.[29] To make one, we take two of the half-silvered mirrors from Figures 2.3 and 2.4 and arrange them as in Figure 2.5. As before, first, we'll show the classical Fresnel wave

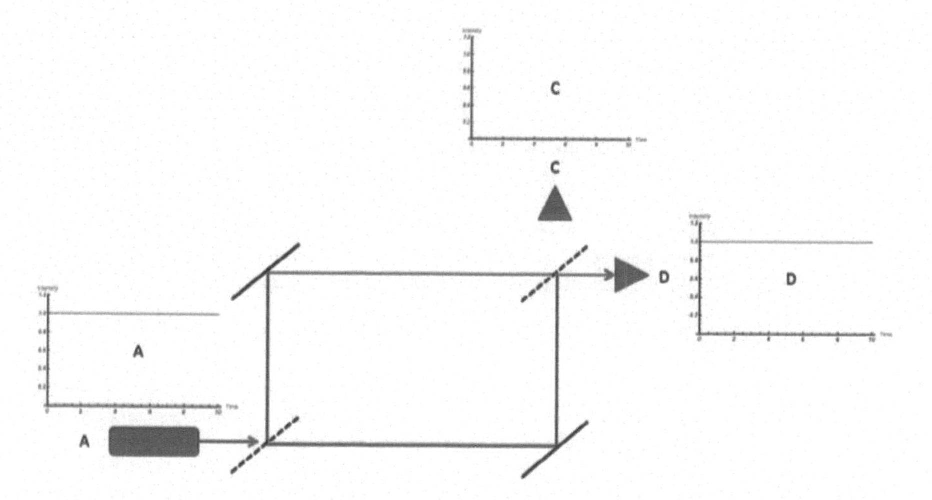

FIGURE 2.5 This is a setup for a classical Mach–Zehnder interferometer. The laser (A) is producing a conventional beam of light of intensity one. The two dashed lines are 50–50 beam splitters. The two solid lines are ordinary mirrors. The upper and lower paths inside the rectangle are balanced to be the same length. As the graphs show, no light at all reaches detector C, and all the light from the laser reaches detector D. We can only explain this experiment with the Fresnel wave theory, which predicts the waves constructively interfere at the output D and destructively interfere at the output C.

29 For more on the Mach-Zehnder interferometer see https://en.wikipedia.org/wiki/Mach%E2%80%93Zehnder_interferometer.

experiment and theory, and then we'll prove the quantum photon experiment and theory. The goal is to see that neither Newton's particle theory nor the Fresnel wave theory can explain all of the upcoming experiments.

In Figure 2.5, we show a classical light-wave Mach–Zehnder interferometer experiment. The interferometer consists of two 50–50 beam splitters (dashed lines) and two ordinary mirrors (solid lines) placed on the corner of a rectangle in such a way that the laser light can take either the upper path or the lower path. Both routes are balanced to be the same length. What we observe in the experiment is that all the laser light goes to detector D. In contrast, none of the light goes to detector C. The Newton particle theory would predict half the particles would come out at *C* and half at *D*. However, the Fresnel wave theory agrees with this experiment. The waves inside the interferometer interfere destructively at C. However, the waves interfere constructively at D and add up to the original laser intensity. Just to show you that this is a real thing, here in Figure 2.6 is a schematic of the device from Mach's paper on the interferometer.

Now we take the classical experiment in Figure 2.5 and perform the quantum version. We keep the interferometer but replace the laser pointer with the single-photon gun used in Figure 2.4. We show the result in Figure 2.7. The photons from the single-photon gun enter the interferometer – bang-bang-bang-bang. Then, as in the classical version of Figure 2.5, all the photons come out to detector D – click-click-click-click – and none ever come out to detector C – blank-blank-blank-blank. The fact that the gun goes bang means that each photon is a particle at the input. The fact that detector D goes click means that the photon is a particle at the output. But for destructive interference to occur at the second beam splitter, so that no photon exits to detector C, the photon must be a wave inside the interferometer – Newton's particles do not exhibit interference – only Fresnel's waves do. There is no crosstalk between photons. If the gun fires one photon per second, it takes about only 30 nanoseconds to go through an interferometer the size of the one in Figure 2.6. There is an eternity between photons, and no two are ever in the device at the same time. Newton, Huygens, and Fresnel got the argument all wrong. They spent years arguing that light is either

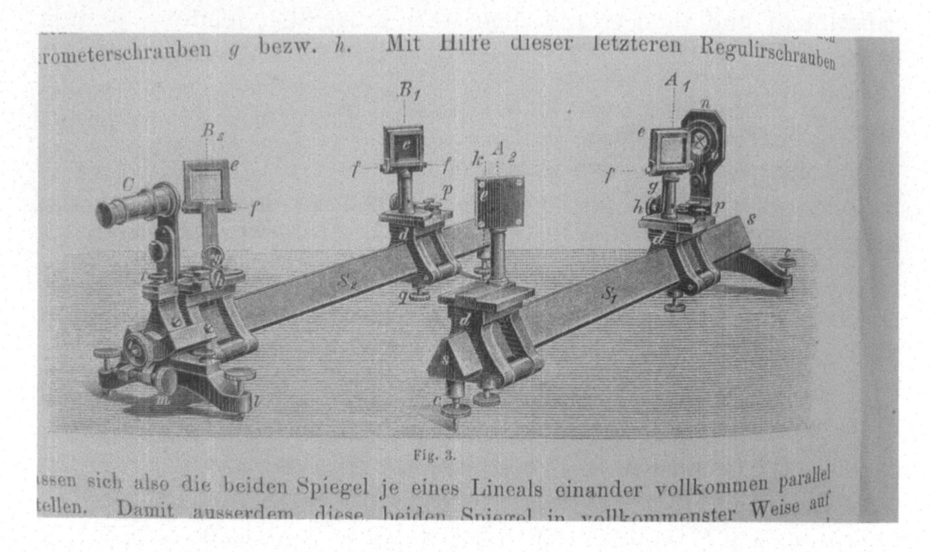

FIGURE 2.6 A detailed line drawing of the Mach–Zehnder interferometer. The four squares are the two beam splitters and two mirrors, and they are mounted on a track so that they can be moved and carefully aligned. Here, A_1 and B_2 are the beam splitters, B_1 and A_2 are the mirrors, and C is the traditional light source. Photo courtesy of Charles Clark, National Institutes of Standards and Technology, taken 20 February 2017 from the actual print version of Mach's paper that he found in the Radcliff Science Library.[30]

a particle *or* a wave. The fundamental unit of light, the photon, is neither a particle nor a wave. This is a bit like Schrödinger's cat – neither dead nor alive.[31]

---

30 The original article is, Zehnder, Ludwig. "A New Interference Refractor." *Journal for Instrumentation News* 11 (1891): 275–285. You can find the entire journal volume scanned in here: https://archive.org/details/zeitschriftfrin11gergoog. Unfortunately, the whole article by Zehnder was torn out of the journal before they scanned it, as pages 275–285 (and only these pages) are missing! Probably they were sold on eBay for a pretty penny. Photo courtesy of Charles Clark. Work is in the public domain.

31 If you don't like that the cat is neither dead nor alive, here is a completely sensible but probably incomprehensible alternative, as proposed by two of my dear friends and colleagues on an exchange on Facebook. I have withheld their last names.

Nick: "… stop telling folks that superposition [of the cat is] 'both states at the same time' – it's not."

Jonathan P. Dowling: "I have to tell them something. In this new book, I'm trying out, "The state of the cat simply does not exist."

Nick: "I hope you don't stop there. Here's what I use for first-year's (freshmen): The quantum state quantifies our ability to make predictions about future measurements involving the

---

particle. I get a tremendous amount of mileage out of this. It cannot be denied in any interpretation of quantum mechanics. And it includes nothing contentious."

Nick: Furthermore, the bra is a yes-or-no question, and the Born rule represents the probability of "yes" to the question given the state (ket).

Jonathan P. Dowling: "Your freshman must hate you."

Nick: "They love it. And they get it. They did well on their assignments and exam using bras and kets."

Jonathan P. Dowling: "I'm writing a book, not teaching a class. I will now mention your approach!"

Nick: "This approach could work for a book since the math is incidental to the concepts. Good luck!"

Jonathan P. Dowling: "It's a popular book. No math. But 'shut up and calculate' does not sell books."

Nick: "It's not "shut up and calculate." It is an acknowledgement that with quantum theory, we have reached the limits of what abstracting a property-filled world outside of ourselves can accomplish (i. e. classical physics). Instead, we must contend with the fact that any properties that are observed about the world are not necessarily there to begin with, even though the world itself is there to be observed and separate from ourselves. We don't do anyone any favours by making up analogies that don't actually work. We can encourage readers to dwell in the "negative space" between concepts that we understand – and in so doing define (or at least outline) a new concept that we do not have a good handle on. This is how intuition is built. This is the most important thing described in the comic you posted [http://smbc-comics.com/comic/the-talk-3]. It is saying that superposition is not "and", it is not "or"; it is instead a way of combining alternatives that are much more subtle and more profound than either of those two classical concepts.

Roman: "Nick, when I try to get philosophical with students (or in more public talks) I use the tetralemma as an analogy and seem to get some mileage from it. Saying that the state is "neither yes nor no" before you make the measurement, etc. https://en.wikipedia.org/wiki/Tetralemma.

Nick: "Roman, with reservations, I think that's a good idea. I'll have to think about it. I tend to come at it from the perspective of well-posed versus ill-posed questions. 'What's the name of your cat?' is ill posed if you don't have a cat."

Jonathan P. Dowling: "I agree, and I do not. No analogy works, but I attest to their usefulness. Wait for the book!"

Roman: "By ill-posed questions, do you mean the counterfactual sort of 'What would the measurement outcome have been if I had actually made the measurement?'"

Nick: "Roman – 'Where *is* the particle?' is ill-posed (in most interpretations) because it assumes that position is ontological. What cannot be denied is that in making a measurement of a particle's position you get an outcome and that if you measure it again immediately afterward you get the same outcome. Anything more than that requires additional assumptions that are false in many interpretations."

Jonathan P. Dowling: (Yep, a popular book written by these two would be a real page turner. This is what nearly all conversations at workshops on the foundations of quantum physics sound like. And recall my two colleagues were trying to advise me how to best explain quantum superpositions to a lay audience. This is why I don't go to workshops on the foundations of physics anymore. I think Prof. Nick's point is that rather than discuss what the wavefunction of a cat "really" is, we should simply say it is a thing that we use to compute probabilities, like dead or alive, since we have no access to what it really is, we should not ask about that. I would suppose a large number of physicists would agree with that point of view, but an even larger number would not. As far as I can tell the tetra-lemma is an old idea that quantum mechanics should be described by a trinary logic. In binary logic, if the cat is not dead then it must be alive. Trinary logic allows for

QUANTUM NONDEMOLITION DERBY[32]

Let's explain this paradox with a story. Alice (A) lives next to a park that she shares with Charlie (C) and Doug (D), who live diagonally on the other side of the park from her. There is a path that runs around the edges of the park that you can take clockwise or counterclockwise. Alice has a pet dog (named Fuzz Lightyear) that she releases every night at midnight out of her front door. It runs into the park at the gate (beam splitter) near A. The park is too dark to see what exactly is going on in there, but a minute later, the dog always emerges into Doug's yard and never into Charlie's yard. Charlie and Doug find this behavior of the

an excluded middle – if the cat is not dead then there is another option besides it must be alive. What that option is depends on who you talk to, but most would say that before the measurement that the cat is neither dead nor alive. See Jammer, Max. *The Philosophy of Quantum Mechanics*. John Wiley & Sons, 1974. http://www.worldcat.org/oclc/488874597.)

Nick: Why insult us for the failure to do something we're not trying to do? If I were writing a popular book, I would not use the terminology I've used above, and neither should you. The fact of the matter is that the wave function is a calculational tool, but it is more than a calculational tool in the same way that a probability distribution is*: it represents our certainty about future measurements. (*Note: this analogy is not for popular consumption because we don't want folks to believe that wave functions merely represent probabilities since there exists a reasonable interpretation of classical probabilities as ignorance of the real state of affairs.)

Jonathan P. Dowling: "Nick not intended to insult. I blame my Asperger's. You have a very well-argued personal opinion that is nevertheless not universally shared. I'm older, but not wiser. And if you plan to attend a workshop on quantum foundations, then you really need to grow a thicker skin. I have knife wounds until this very day!"

Nick: "Apology accepted, and your self-awareness applauded. Insult ≠ injury! I'm not hurt. I'm pointing out the impropriety of the insult for the circumstances. I've attended many a foundations workshop. I revel in the debate."

Jonathan P. Dowling: "Nick, if only I could tell."

Roman: "Jonathan, what is your personal opinion on these matters of presentation of the ontic and/or epistemic nature of quantum states? We all know that much has been said in the past; but where are we now in the discourse? Your "The state does not exist" sounds like an overly zealous ontological statement to me – there must be more you're thinking of!

Nick: "Roman, this is exactly why I'm advocating for the 'least necessary claim' about the quantum state. Saying it 'doesn't exist' is a claim about ontology and worth avoiding."

Jonathan P. Dowling: "I like Nick's point that asking you the name of your cat makes no sense if you don't own a cat. I'm going to say *neither* dead *nor* alive. *Something* is in the box."

Roman: "It's more-or-less clear that non-existent things don't have attributes. (If I don't own a cat, then my cat has no attributes.) More interesting are the attributes of existent things, especially before they are measured. I like Nick's statement about making the least necessary claim. I also like the QBists' approach of talking about betting strategies and personal knowledge."

32 My colleague, Chris Monroe, hates the term "quantum nondemolition," and so I'm deliberately out to goad him. See https://physicstoday.scitation.org/doi/10.1063/1.3541926.

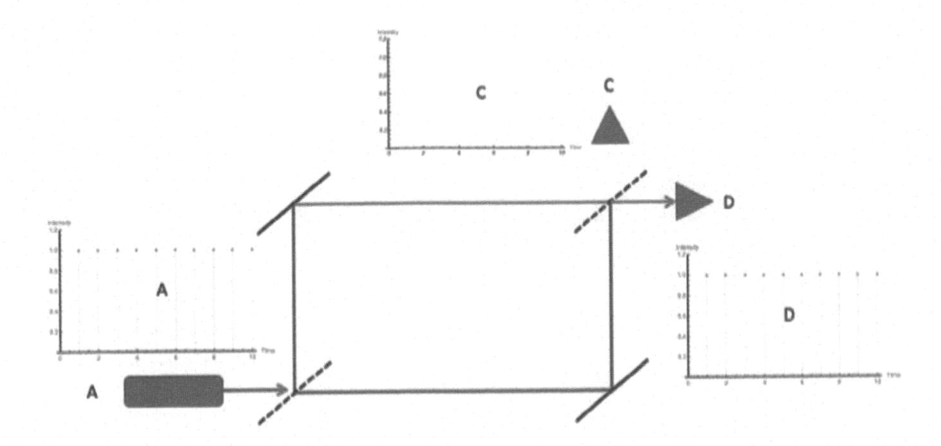

FIGURE 2.7 A single-photon gun A fires one photon per second into the interferometer. Just as in Figure 2.5, all the photons come out to detector D and none ever come out to detector C. The fact that the photons never exit to C is due to their wave-like property of destructive interference. The fact that the photons hitting detector D go "click-click-click-click" is due to their particle-like property.

animal odd, as they have identical yards. Why does the dog never run into Charlie's yard? They figure Alice has trained the dog to always run into Doug's yard, which is a bit odd as the two paths around the park are identical. But Alice could instruct Fuzz to still go straight at the first gate and always make a right turn at the second gate into D (or make a left corner at the first gate and then go straight into D). Everything seems sensible, and they conclude that Alice is not a very good neighbor, nor a very responsible dog owner, but a little bit crazy, with a genius dog.

In Figure 2.8, we show a slightly modified experiment. In the upper arm of the interferometer, we place a quantum nondemolition detector (QND). The QND, as the name suggests, records whether the photon took the upper arm (or not) in each 1-s time interval, but without destroying the photon (as the triangle-shaped detectors C and D will do). It records these measurements on a USB flash drive, with zero for no photon went this way, and one for one photon went this way. This seemingly innocuous device *completely* changes the outcome of the experiment. We rerun the test with the QND in place – *without ever looking at the results on the QND's flash drive!* What we see in Figure 2.8

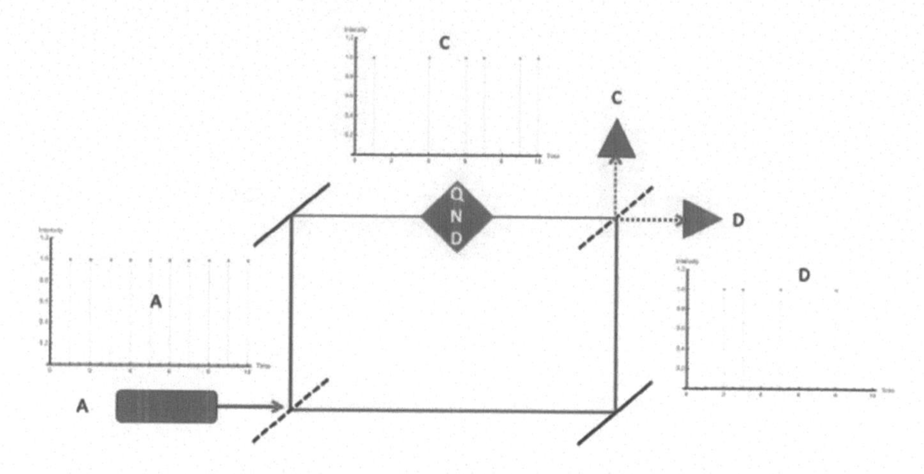

FIGURE 2.8 In the upper arm of the interferometer, we place a QND, which records if the photon took the upper arm or not. The QND spoils the wave interference, and the photons emerge randomly with a 50–50 probability out to C or D as in Figure 2.4, showing their particle nature. Note that this outcome occurs for runs of the experiment when the QND records nothing at all. The interference is destroyed even if the photon takes the lower path and goes nowhere near the QND.

is that we are back to the photons popping out randomly to the D or C detectors, with a 50–50 probability, as in Figure 2.4. The mere presence of the QND device, even if nobody is reading its output of zeros and ones, has destroyed the wave interference, forcing the photon to always behave like a particle. Even better – we can rig the QND to turn on or off randomly, long after the photon has left the source, and still, we get the same result. That is, *nothing in the single-photon gun itself* can be the cause of this new outcome. If the QND is off, the photon always clicks at detector D (indicating the wave nature of the photon). When the QND is on, the photon clicks at either C or D with a 50–50 probability (indicating the particle nature). What is bizarre is that the interference is spoilt even when the QND detects nothing. That is, presumably, because the part of the photon that sets off the detector took the lower path. *How then did that photon on the lower path even know about the QND in the upper path?* The interferometer could be lightyears across, and still, the interference is spoiled if the QND is there – lightyears away from the path the photon "actually" took.

Back to Fuzz lightyear. In the new set up, Charlie and Doug have installed a motion sensor on the upper northern fence of the park. The motion sensor is the QND. The sensor records on a USB drive if something runs by the wall. They can turn the sensor on or off with a remote control. They do a series of experiments. Every night at midnight, they flip a coin, and if heads, they turn the motion sensor on, and if tails, they leave it off. On nights when they get tails, and it is off, like before, Fuzz always runs into Doug's yard, and never into Charlie's yard, as in Figure 2.7. However, on nights when the coin is heads, and they turn the motion sensor on, Fuzz randomly runs into either Doug's yard or Charlie's yard, with a 50–50 chance, unrelated to the coin toss, as in Figure 2.8. But here is the weird part – Charlie and Doug observe this 50–50 erratic behavior – *even on nights when the motion sensor is on but does not detect anything.* (Which would seem to imply that the dog took the lower path and did not go near the motion sensor at all.)

How do they interpret this behavior? One obvious solution is that Alice has two identical dogs, Fuzz and Scuzz, and Alice trains Fuzz to always run into Doug's yard, and Scuzz is untrained and randomly runs into either yard. Alice must also have a telescope and can see the little blinking red light on the motion sensor, so she knows when it is on or off. If the motion sensor is on, she releases random Scuzz, and if the motion sensor is off, she releases well-trained Fuzz. This telescope would also explain that they get Scuzz even if the motion sensor does not detect him. Alice is preparing which dog to send based on her observation of the motion sensor's on-ness or off-ness.

To thwart her little scheme, instead of using a remote control, Doug and Charlie upgrade the motion sensor so that it turns itself on about 3 ns before 12:30 AM. It then turns itself off about 3 ns after 12:30 AM so that it is only active when the dog (potentially) runs by. Additionally, they hook it up to a random-number generator, so that it has a 50% chance of being on and a 50% chance of being off each night when the dog runs through the park. They observe precisely the same behavior as before – on nights when the motion sensor is altogether off the dog always runs into Doug's yard – on nights when it is on for 6 ns, the dog randomly runs into either Charlie's yard or Doug's yard with a 50–50 chance. But how can they explain this? Light only travels about 1 m in 3 ns. There is no way that Alice, looking at the motion sensor with her telescope, can tell whether it will be on or off at 12:30 AM, as the light-travel time will cause her to see the on-off light only after the motion sensor has (or has not activated) and the dog has long passed (or not passed) by it.

That is Alice cannot *choose* to send Fuzz when the detector is on (or Scuzz when it is off) since she always releases them precisely at midnight, and according to the laws of physics, she cannot know whether the motion sensor will be on or off at 12:30 AM when the dog is halfway through the park. The only interpretation is that *Alice* does not choose which dog to send through the yard. Rather – *the yard* chooses which dog will run through it! Even weirder, Charlie and Doug always get random Scuzz on nights the motion sensor is on – even when it detects nothing – nights when Scuzz must have taken the lower path. The mere fact that the sensor is on – *even if it identifies nothing!* – is enough to collapse the wave function of the dog from well-trained Fuzz into random Scuzz, although the dog(s) never went near the thing. "How often have I said to you that when you have eliminated the impossible, whatever remains, however improbable, must be the truth?" – Sherlock Holmes.[33] Since it is impossible for the signal to reach Alice going faster than the speed of light, she can't predetermine the outcome of the run.

No matter how improbable it may seem, it is the *park* – not Alice! – that determines which dog, Scuzz or Fuzz, will run through it.

This improbable performance is precisely the behavior we see with photons in a lab. The QND in Figure 2.8 can be programmed to turn on or off at random when the photon passes by. If on, then there is a 50–50 chance for C or D to get a click. If off, then only D receives a click. The first result reveals the wave nature of the photon, as only interference explains why it always goes to detector D. A click at C conclusively shows the particle nature of the photon. The sensor C gets a click only if the photon acts like a ping-pong ball. Alice's single-photon-source could be lightyears away, so Alice has no possibility of choosing the wave or particle nature. The experiment, the interferometer controlled by Charlie and Doug, determines the fundamental essence of the photon. Charlie and Doug can decide, long after the photon has left Alice, if it will be a wave or a particle. Also, the mere

33 Doyle, Arthur Conan. "The Sign of the Four." *Lippincott's Monthly Magazine*, Ch.6, 11. Spencer Blackett, 1980.

presence of the QND in the experiment collapses the photon into a particle – even if the photon goes nowhere near the QND. Welcome to quantum unreality. The photon has no wave or particle nature until it is measured. Before that happens, the photon's character is unknown and unknowable. This experiment is a modified version of something that is called a delayed choice.[34] There are not two dogs, Fuzz and Scuzz, but there is just one dog. Its Fuzziness or Scuzziness is revealed only through an experiment. Before that observation, it is neither Fuzz nor Scuzz – it does not exist! This lack of existence is what we mean by quantum unreality. Quantum objects do not have a reality independent of the experiment and observers. As Einstein once, plaintively, complained to a colleague, "Do you really believe the moon is not there when you are not looking at it?"[35] In our scenario, we have to believe that Alice's animal is neither Scuzz nor Fuzz until Charlie and Doug set up an experiment to look for it.

VAIDMAN – DUDE'S THE BOMB!

Let's play one last game with a Mach–Zehnder interferometer – this one's a doozy! It is a variant of the above experiments that teases out one last bit of bizarreness about quantum states of photons. These quantum-thought experiments seem to be more popular, the more violent their nature. Take the case of Schrödinger's cat, where half the time, he poisons the poor feline to death in a cage.[36] The experiment I would like to describe next is called the Elitzur–Vaidman bomb tester.[37] I first heard of this idea at a conference in Maryland in 1993 during a talk by the quantum iconoclast

34 See https://en.wikipedia.org/wiki/Wheeler%27s_delayed_choice_experiment. See also our recent paper, Huang, H.-L., Y.-H. Luo, B. Bai, Y.-H. Deng, H. Wang, H.-S. Zhong, Y.-Q. Nie, W.-H. Jiang, X.-L. Wang, J. Zhang, Li Li, Nai-Le Liu, Tim Byrnes, J. P. Dowling, Chao-Yang Lu, Jian-Wei Pan. "A Loophole-Free Wheeler-Delayed-Choice Experiment." https://arxiv.org/abs/1806.00156. See also, https://www.quantamagazine.org/closed-loophole-confirms-the-unreality-of-the-quantum-world-20180725/.

35 Bernstein, Jeremy. "Einstein: An Exchange." *The New York Review of Books*, August 16, 2007. http://www.nybooks.com/articles/2007/08/16/einstein-an-exchange/.

36 Heisenberg and Schrödinger are driving down the freeway when a policewoman pulls them over. The policewoman says to Heisenberg, "Do you know you were doing *exactly* 100 miles per hour?" Heisenberg exclaims, "Great! Now we're *totally* lost!" The policewoman says, "Okay wise guys, I need to search the vehicle – pop open the trunk!" The officer looks in the trunk and yells back, "Do you two jokers know you have a dead cat back in here?" Schrödinger throws up his hands and yells, "Well – *now* we do!"

37 For more on the bomb tester, see https://en.wikipedia.org/wiki/Elitzur%E2%80%93Vaidman_bomb_tester#CITEREFElitzur_Vaidman1993.

Lev Vaidman himself. Vaidman, an Israeli physicist, has a reputation for giving a speech where he draws some simple interferometers, as I have sketched above, makes a few calculations, and reaches a conclusion that causes the audience to erupt into a brawl. During this talk on the bomb tester, distinguished professors were screaming that he had to be wrong. I was a mere postdoctoral intern at the time, so instead of immediately screaming, I rapidly carried out the calculation myself in my notebook as the room raged. Then I stood up and yelled, "I have just confirmed the calculation – Vaidman is right!" More chaos ensued, followed by fisticuffs, but indeed Vaidman *was* right. The trick is to replace the QND in Figure 2.8 with a quantum *demolition* detector.

As told by Vaidman, in the experimental setup in Figure 2.7, randomly from run-to-run, a malicious Dr. Evil secretly places a single-photon hair-triggered bomb in the upper path. If even a single photon strikes the weapon, causing its photodetector to click, the gadget explodes, destroying the lab and killing everyone. The game is to see if the experimentalists Alice, Charlie, and Doug can detect the existence of the bomb without setting it off. According to classical Newtonian mechanics, they cannot, since to expose the bomb they have to interact with it, such as shining a photon on it, immediately setting it off. The weirdness of quantum theory allows for a way out.

If the bomb is not there, then we have precisely the setup of Figure 2.7, where the photons always exit to Doug's detector D and never to Charlie's detector C, due to quantum interference. If the bomb is there, then we have a modified version of Figure 2.8, where the photons randomly exit to C or D, but there are only half as many, as 50% of the photons take the upper path and hit the bomb and go no further. The game is that of a very risky quantum Russian roulette. Alice fires a single photon into the interferometer. The presence of the bomb destroys the quantum interference, just like the QND, and the photon now behaves like a ping-pong ball. After the first beam splitter, it has a 50% chance of taking the upper path, hitting the bomb and setting it off, killing everybody.

On the other hand, it has a 50% chance of traversing the lower route. If it does that, it has a 50–50 chance of going to either C or D. Therefore, it has a 25% chance of being detected at C and another 25% chance of being caught at D. If we find it at D, we do not know if there is a bomb or not, for that is the outcome we would expect for no explosive at all. However, if detector C clicks, then we know for sure that the bomb is there – without setting it off! Only the presence of

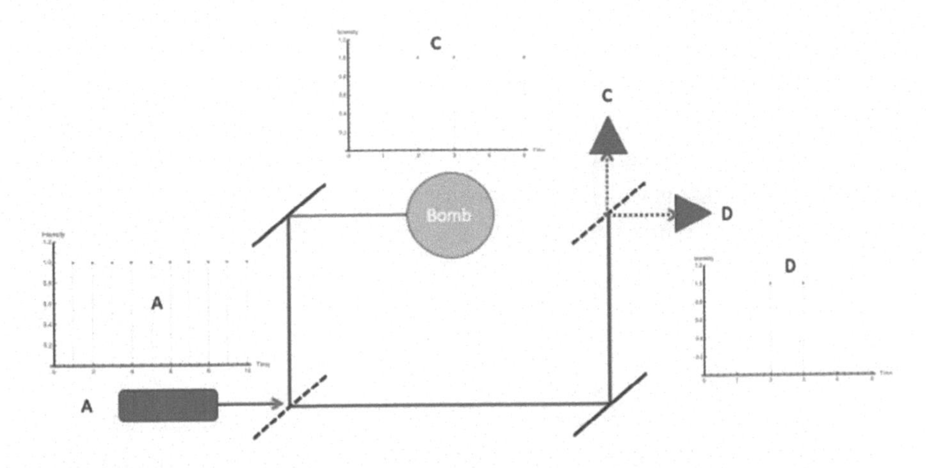

FIGURE 2.9 The Elitzur–Vaidman Bomb Test. The source A shoots a single photon into the interferometer. If the bomb is absent, the photon always exits to detector D, which records a click due to quantum interference, as in Figure 2.5. If the bomb is present, then it destroys the interference, and there is a 25% chance the photon will exit and strike detector C, which tells us conclusively that the bomb is there. If the photon exits to detector D, which happens 25% of the time, we learn nothing. If neither detector C nor D records a click, then the bomb explodes, killing everybody in the lab, presumably because the photon took the upper path and set off the bomb.

a bomb would spoil the interference and allow the photon to be detected at C. To summarize – we have a 50% chance of blowing ourselves up, a 25% chance of learning nothing, and a 25% chance of discovering that the bomb is there without setting it off.

There we have it. Using quantum interference, 25% of the time, we can detect the presence of a bomb with a single photon whose energy and particle-ness *went nowhere near the bomb at all.* The electrical power of the photon is what the detector detects. The bomb could sit on the upper path – lightyears away from the lower trail that "the photon" *must* have taken to arrive at detector C unscathed. But if the energy of the photon went nowhere near the bomb, and never interacted with the bomb, *how on earth did that photon ever know that far-distant explosive was even there?* If you believe in the physical reality of the photon, the only way the photon could ever exit to detector C is if the bomb is present. In that scenario, you might conclude the photon had to take the lower path to avoid setting off the bomb. But what then did the bomb disturb in the upper arm of the interferometer to destroy the interference, if the "photon" (or at least everything we'd want to call a bullet of energy) went nowhere near it? I don't have a good answer as there is none. The photon has no objective reality independent of the experiment. If you like, I can tell you that most of what we think of as a photon in Newton's terms – a speedy bullet of light energy – took the lower path and avoided setting off the bomb in the top way. However, some ghostly Huygens–Fresnel wave-like essence of the photon – whatever the heck that means – took the high path to detect the bomb's presence without carrying enough of whatever makes up the rest of the photon to set the bomb off.

To summarize the whole point of this chapter, we have learned that quantum-mechanical objects like photons have no objective reality like a bullet does or a wave does. They do not obey common sense. But common sense is our ability to predict the future based on our past experiences, and we have no prior experiences with quantum objects like photons. If we were tiny quantum-mechanical beings that observed this kind of behavior all the time, then it would make common sense, but we are not such beings. We write down our equations and see these clicks at the detectors in the lab, and we have to embrace the weirdness, learn how to use it, and then put it to work. Asking what the photon is "really" doing in the interferometer is a fool's errand. Niels Bohr, a founder of quantum theory, called this particle-wave weirdness quantum duality. He took the stance that our classical concepts such as wave or particle are insufficient to describe the quantum world. Others, like Albert Einstein and Erwin Schrödinger, wanted to know what the photon was "really" doing, but they failed to get a satisfactory answer. Another point of view is that of Werner Heisenberg, who took a minimalist view – we have a theory, we calculate stuff, and if our calculations agree with what we measure in the lab, then that's all we can ask for. But humans strive for meaning. Until this day, physicists still argue about what the photon is "really" doing in that interferometer.

THE UNDETECTABLE QUANTUM TRIPWIRE

To make this bomb tester even stranger, we can do the following. Instead of 50–50 beam splitters, we can adjust them so that they are 99.9999 … percent transmissive and 0.000…00001 percent reflective. Also, we can make nested Mach–Zehnder interferometers and allow the photon to go around many times in multiple attempts to detect the bomb. The setup is a bit complicated, but for all practical purposes, the result is that of Figure 2.10. We tune the interferometer so that none of what we would like to call a photon ever takes the upper path, and its energy and polarization and everything else we'd want to call a photon always takes the lower way.[38] You might think there is no way to detect the bomb at all, since nothing is ever "really" in the upper path. All the photons we send in, come out the other side and hence we never set off the bomb. But that ghostly wave part of the photon – which does not carry energy or anything else tangible – still drifts spectrally through the upper path to cause interference. If the bomb is not present, the quantum interference forces all the photons out to detector D, and if the bomb is present, then it destroys the ghostly interference, and the photons always come out C. We never blow up the lab!

A click at detector D always means no bomb (Figure 2.10a), and a tick at detector C always means yes bomb (Figure 2.10b). If we associate the entire notion of "the photon" to be something that causes a detector to click, then we must assume that the photon always takes the lower path. But then how does the photon know that there is a bomb in the upper path – possibly lightyears away? The only conclusion is that a photon is something more than that-which-makes-a-detector click. Again, if you like, there is some ghostly apparition of the photon that interrogates the upper path to decide if the bomb is there or not, but that ghost of the photon does not carry energy or anything else that would set off a bomb.

To pedal back a bit from the bomb, we discuss an undetectable quantum tripwire. A tripwire is a wire that you run around your campground with a bell attached to it. If a bear tries to sneak into your campsite at night to eat you, it would trip over the wire, ring the bell, and hopefully wake you up in time to defend yourself.

In the more modern version of this, you've seen these things in action-adventure spy-thriller films such as *Mission Impossible*.[39] Tom Cruise is trying to sneak into some secure facility to steal a computer

38 Anisimov, Petr M., Daniel J. Lum, S. Blane McCracken, Hwang Lee, and Jonathan P. Dowling. "An Invisible Quantum Tripwire." *New Journal of Physics* 12 (August 2010). https://doi.org/10.1088/1367-2630/12/8/083012. A free online version is here https://arxiv.org/abs/1002.3362.

39 Make your visible classical tripwire, a la mission impossible, following the instructions of this video: https://www.youtube.com/watch?v=jxxY9uTAD6I.

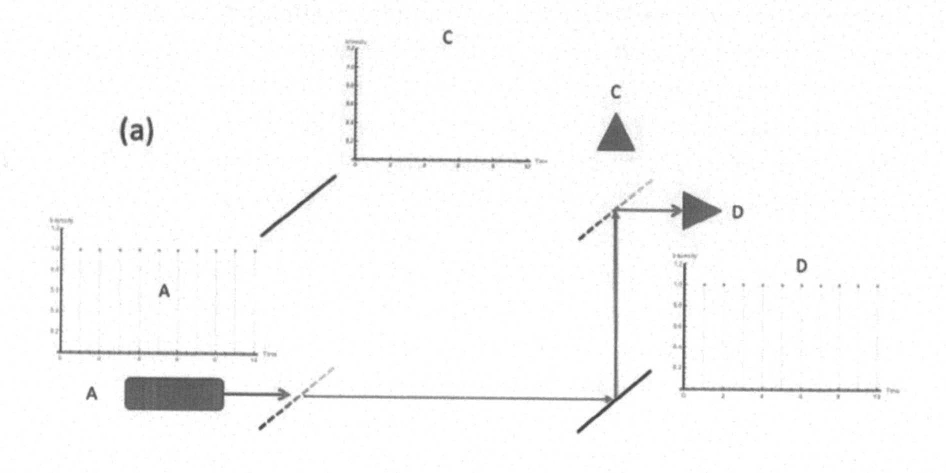

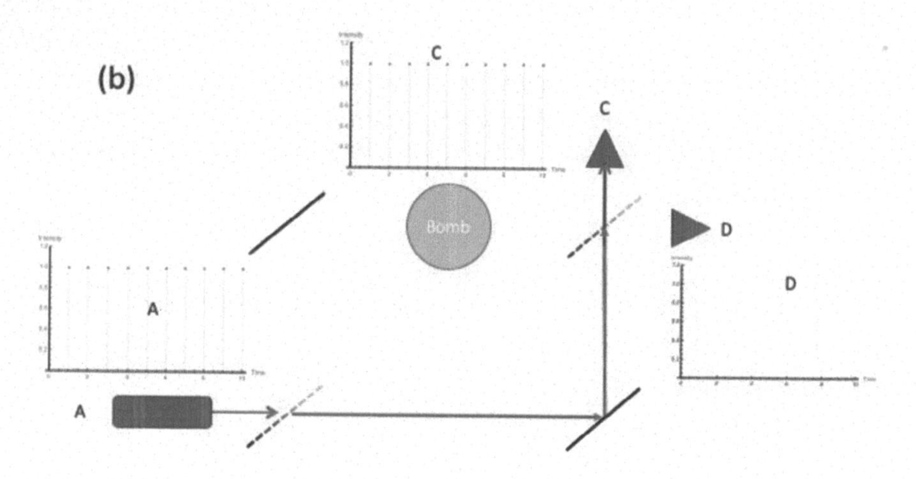

FIGURE 2.10 The new and improved bomb tester. We adjust the beam splitters and paths in such a way that if the bomb is not present (a), then the detector D always fires, and C never fires. If the bomb is present, then C always fires, and D never does. The classical description of this device is that all of the light wave takes the lower path. But then how can the light wave know about the bomb? The quantum description of the device is that the energy-carrying ping-pong-ball nature of the photon – always takes the lower path. Nevertheless, the nonenergy-carrying undetectable wave-like essence of the photon takes the top way, providing constructive or destructive interference.

disk, and he is climbing through an airshaft when he suddenly sees all these green laser beams crisscrossing in front of him. He knows that if any part of his body blocks the beam, it will set off an alarm. Whoever designed this tripwire should be taken out and shot. First of all, why is the air shaft a square meter across to allow people to climb around in it in the first place? Second of all, if you want intruders to set off the alarm, why make the laser beam bright green so the intruder can easily see it? Tom Cruise sees the laser beams and effortlessly shimmies and slithers and limbos between them without blocking the rays and setting off the alarm.

If I were designing this contraption, I would first think of using an infrared laser beam, which is invisible to the naked eye. That way, Cruise would run into the beam without knowing it was there. But you can get night goggles that can see in the infrared or this dust you can blow down the shaft that glows in infrared, so you can still look at the beams and thus thwart the tripwire. However, detecting the laser beam means somehow witnessing the energy of photons in the shaft and then navigating around them. What if there were no energy of the photons in the beam at all? The improved bomb tester becomes an undetectable tripwire. In Figure 2.10, instead of a bomb, we have Tom Cruise. When the system is armed, all the photons exit to detector D. But if Cruise crosses the upper beam, then the photons immediately exit to detector C, and Cruise sets the alarm off. There is no way Cruise can "see" the tripwire because there is no energy of the photon to see. Rather, there is only the undetectable ghostly wavelike essence of the photons. The only way to detect that essence is to block the beam, thereby setting off the alarm. The tripwire is not only invisible, but it is also undetectable, and it is not there in any real sense of the word "real".

"Last night I saw upon the stair, a little man who wasn't there. He wasn't there again today. Oh, how I wish he'd go away …."[40]

EXTRAORDINARY POPULAR DELUSIONS AND THE MADNESS OF CLOUDS

We can use the improved bomb tester to make an uncanny imaging device – a camera that takes photos in pitch darkness where there is no light at all. From chapter one, we know that cameras collect reflected

40 From the poem "Antigonish," by William H. Mearns (1899). It is from a play about ghosts.

and scattered light off of an object and focus the light onto a digital array of camera pixels and thus form an image for my Pixel XL smartphone. If I stand in a perfect, pitch-black room and turn off the flash, then no matter how long I leave the shutter open, the camera of the Pixel XL will never get a photo of anything. Let me introduce the Quantum Pixel XW, where XW stands for eXtra Whacko. In Figure 2.11, we replace the bomb with an image. The image is a black and white (but not grayscale) photo of a cat that we print onto a sheet of transparent plastic. The film of plastic and the upper arm of the interferometer are enclosed in a completely dark room with just two pinholes to allow the beam containing no photon energy to pass.

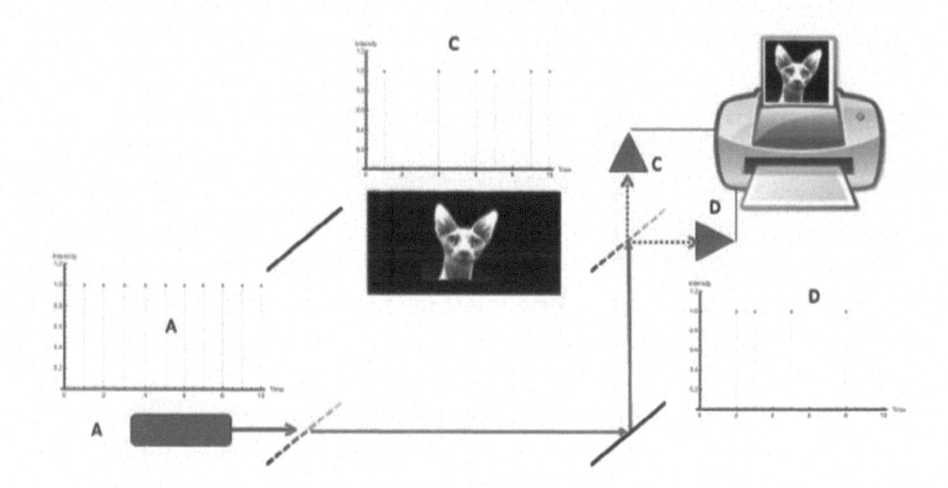

FIGURE 2.11 Quantum seeing in the dark. The improved bomb tester is an imaging device. All of the energy-carrying parts of the photon take the lower path. The nonenergy-carrying wave-like spooky essence of the photon takes the upper path and goes into a pinhole punched in the side of a completely black room and then comes out a pinhole on the other side. The inside of the box is entirely dark. A transparency with a black and white image of a cat hangs in the room attached to a contraption that raster scans it pixel-by-pixel in the upper beam. When the cat pixel is transparent, the beam is unblocked, and outside the room, detector D always fires. When the cat pixel is opaque, it blocks the beam, and detector C then fires. The detectors are attached to a printer printing a black square when C fires and white square when D fires, thus reconstructing the image of the cat.

The cat image is composed initially of black and white pixels, but after transferring to the plastic sheet, the pixels are either opaque (black) or transparent (white). We slowly raster scan the image of the cat through the upper arm of the interferometer, pixel by pixel, in a square grid. For each pixel, we fire a photon into the interferometer. If the pixel is transparent, that is equivalent to no bomb, and the ethereal and undetectable essence of the photon in the upper arm is allowed to pass and combine with the rest of the ping-pong ball nature of the photon in the lower arm at the second beam splitter. We hook the two detectors to an old dot-matrix printer that reconstructs the image (here that of a hairless cat), pixel by pixel, on a sheet of paper. If detector C clicks, that corresponds to a transparent pixel on the transparency (no bomb), and the printer prints a white square pixel on the paper. If detector D clicks, that corresponds to an opaque pixel on the transparency, and the printer prints a black square. Eventually, we have a lovely image of the cat that is identical to that on the original slide. We have taken this image of the cat in a completely dark room with no light in it whatsoever and no photon energy from the interferometer whatsoever. The undetectable wave-like nature of the photon samples the slide. Exploiting that feature, we construct the image with nothing our eyeballs would see as light in the room at all. The experiment is somewhat hard to do with a cat but has been carried out imaging such small things as human hairs.[41]

To emphasize again how weird this is, if you were inside the darkened room, you would see nothing. (The human eye is a pretty good single-photon detector.[42]) You could press your eye up to the input pinhole and see nothing. You could insert your eyeball anywhere in the interferometer beam and see nothing. There is no light in the room, as the energy associated with the photon is not in that space. But it is that energy that triggers the rods in your retina to see something. The machine dutifully reconstructs the image of the cat, in complete darkness, occasionally interrupted by your eyeball blocking the beam.

---

41 Kwiat, Paul, Harald Weinfurter, and Anton Zeilinger. "Quantum Seeing in the Dark." *Scientific American* 275, no. 5 (November 1996): 72–78. http://www.jstor.org/stable/24993449.

42 Holmes, Rebecca M., Michelle M. Victora, Ranxiao Frances Wang, and Paul G. Kwiat. "Testing the Limits of Human Vision with Quantum States of Light: Past, Present, and Future Experiments." SPIE Proceedings Volume 10659, Advanced Photon Counting Techniques XII; 1065903, 2018. https://doi.org/10.1117/12.2306092.

What is in the beam taking the photo of the cat? The part of the photon that contains no energy and is undetectable by any means. Sometimes, when referring to quantum particles like a photon, people say it is less like a ping-pong ball and more like a cloud.[43] That does not do it justice. If the photon is a cloud, then all the water droplets of that cloud are in the lower arm, and no water droplets at all are in the upper arm. The upper arm contains the part of a cloud that remains after you remove all the water from it. What part of a cloud is *that?*

TO COMPUTE OR NOT TO COMPUTE – WHAT IS THE QUESTION?

We now take the bomb-tester to its screwball and ultimate limit. In counter-factual quantum computing, we replace the bomb with a complicated interferometer that is an all-optical quantum computer, which we place in the darkened room. The quantum computer is turned on and is awaiting a program from Alice, who controls the photon source at *A*. Alice programs a sequence of zeros and ones as follows: If the bit is a one, she sends a photon. If the bit is a zero, she sends no photons. We clock the whole thing so that a nanosecond separates the time bins. If we replace the beam splitters with mirrors, then all the photons (or not) take the upper path and enter the quantum computer. The computer runs the program and produces a different sequence of zeros and ones, the output, that detector C records. But now replace the mirrors with the different beam splitters as in the improved bomb tester. All the energy-bearing parts of the photons from Alice, containing the input sequence of zeros and ones, take the lower path and never interact with the computer at all. The computer, instead of processing the photons, now blocks or does not block the path according to its program. In that way, it changes Alice's input into the computed output. The detectors C and D record the output of the computer. However, nothing that you would call a photon ever interacted with the computer at all. It was manipulating nothing. And yet the correct output is recorded at C and D.

If we turn off the quantum computer so that it either always blocks the upper path (or always unblocks it), then only D would click, or C would click, respectively. The output would be identical to Alice's

43 Hossenfelder, Sabine. *Lost in Math: How Beauty Leads Physics Astray*, 6. New York: Basic Books, 2018. http://www.worldcat.org/oclc/1005547825.

input, and the computer would *not* have run its program. The computer does not interact with the bit-carrying photons at all, but it still must be turned on to run the program. In case you believe that I have completely lost my mind, please note that physicists have demonstrated a small version of this counter-factual quantum computer in the lab.[44] (Note that this is only a necessary but not sufficient condition to preclude me from having lost my mind.)

EINSTEIN'S BUBBLE – TOIL AND TROUBLE

When my Ph.D. advisor suggested I study a single photon in empty space, I retorted that was too difficult, and spent four years studying empty space. I believe the above sections attest to my wise choice. The purpose of this chapter is to show how bizarre is the quantum mechanical notion of a single photon. I want to round the episode out with a few more examples of the weirdness of single photons before I – gasp! – discuss *two* photons. In the above experiments, I have shown that photons are *unreal* – they do not exist independently of the machines we use to observe them. I have also shown that they are *uncertain* – measurements on a photon can produce completely unpredictable outcomes. Here I would like to make the case that the photon is nonlocal more strongly – that is, the photon can have a wave-like existence spread over lightyears, but a particle-like existence is concentrated at a single detector. In one of his early attacks on quantum theory, Einstein proposed the paradox of the bubble. In Figure 2.12, we show a sphere that is one lightyear in radius. At the surface of the sphere are billions of astronauts with single-photon resolving eyeballs peering into the shell. At the center of the sphere is Alice (A) with her single-photon source. The source is adjusted, according to the wave interpretation, to emit a single photon in a perfectly spherical wave.

According to the quantum-wave interpretation, the single photon spreads like an ever-increasing soap bubble, getting thinner and thinner and thinner until – at the moment the balloon hits the sphere of astronauts – it pops. According to the quantum interpretation, at that instant, all the energy of the photon, spread thinly over the bubble, races instantaneously to the eyeball of one – and only one – of the astronauts, where she sees the *entire* single photon. Photons are

---

44 Dowling, Jonathan P. "Quantum Information: To Compute or Not to Compute?" *Nature* 439 (23 February 2006): 919–920. https://www.nature.com/articles/439919a.

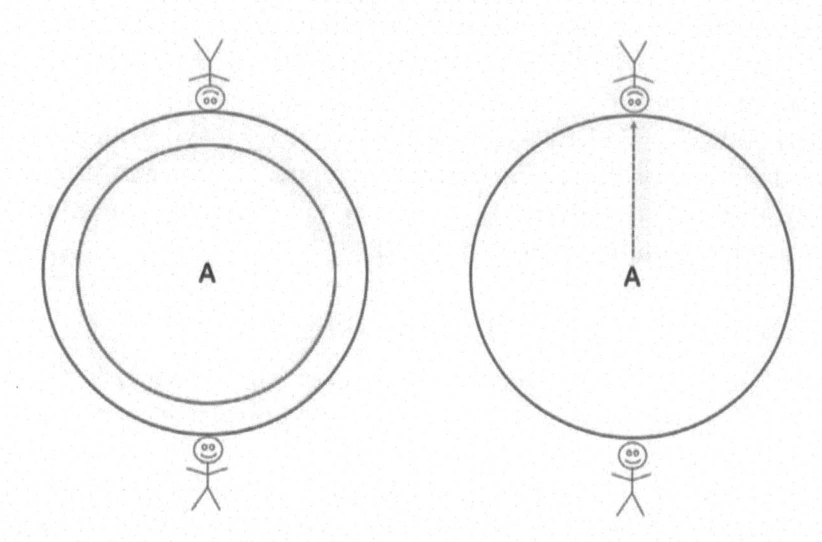

FIGURE 2.12 Einstein's bubble paradox. Alice (A) emits a single photon inside a sphere of single-photon-detecting astronauts – two of which we show on the North and South Poles. On the left, the photon expands out like a growing soap bubble. If the North Pole astronaut sees the photon, the bubble pops, and quantum theory says that she sees all of the photon and the South Pole astronaut sees nothing. Somehow the entire remnants of the popped balloon instantaneously zoom to the North Pole, and somehow the eyeball at the South Pole knows not to see anything. To the right is what the classical hidden variable predicts. Alice's photon source fires the photon-like bullet in a completely random direction. The shot travels along a single path, not a bubble, but you don't know which way. If the North Pole astronaut gets the bullet, then the South Pole gets nothing. The results of the left and right explanations are identical, so why not adopt the more straightforward hidden-variable theory on the right?

indivisible! Let's say she is at the North Pole in the figure. Classically, this photon spreads so thin that no astronaut would ever see it, as its spread-out energy would be too weak to trigger anybody's eyeball. But quantum theory says that the photon randomly triggers one and only one eyeball, and when that eyeball is triggered, no other eyeball is triggered.[45] How does the nontriggered eyeball of the other astronaut at the South Pole, two lightyears away, *know* not to see the photon?

---

45 Musser, George. *Spooky Action at a Distance: The Phenomenon that Reimagines Space and Time*, 88. New York: Farrar, Straus, and Giroux, 2016. http://www.worldcat.org/oclc/965513480.

This is the paradox. The wave interpretation says the photon is spread over a bubble a lightyear in radius. Upon impact with the sphere of detecting eyeballs, the photon collapses randomly to just one eyeball, and the other eyeballs see nothing. If the energy of the photon is spread over the entire bubble, then quantum theory predicts that this energy all rushes instantaneously to the seeing eyeball. In the worst-case scenario, the bubble-thin energy on the opposite side of the sphere from the detecting eyeball, needs to move instantaneously a distance of two lightyears to get to the all-seeing astronaut in time for all of the photon to be seen. Einstein did not like that! In his special theory of relativity, he provided evidence that nothing can move faster than the speed of light. If quantum theory is correct, then the bubble energy of the photon moves a maximum distance of two lightyears in zero time – infinitely faster than the speed of light. Is there any way out? Yes, there is, but the way out forces us to give up the classical theory and the quantum theory, and to embrace something nonquantum and entirely new.

We completely resolve the bubble paradox in the following manner. Suppose that Alice's photon source is not a spherical wave at all but a machine gun that fires photons randomly in all directions in an unpredictable pattern. In the classical particle theory, this would be fine, except it would always be possible in principle, if one calculated the motion of the machine gun, to predict in which direction the photon flies. In the new and improved classical theory, that information is hidden from us because we don't have access to the machine gun's firing pattern. This new theory is neither classical nor quantum theory but rather a hidden-variable theory. Here the hidden variable is the random direction the gun is firing, which we either do not or cannot know. This hidden-variable theory resolves the paradox. Instead of a bubble, we have a photon bullet speeding towards one detector or another at random. The shot contains all the energy of the photon. When it, by chance, strikes the eyeball of one astronaut, it cannot hit the eyeball of another since there is only one bullet. As we repeat this experiment over and over again, it agrees with the quantum theory. Random eyeballs on the sphere see one and only one photon per shot, but after many shots the energy is evenly distributed over the surface of the bubble. The random-firing hidden-variable argument is not classical theory or quantum theory – it is something new. Thence, Einstein launched his counterattack – can we not replace the quantum theory, with its spooky action at a distance and collapsing bubbles, with something sensible like a randomly firing machine gun? The answer is that we cannot, but this paradox is not sufficient to tell us why. That explanation will have to wait until the next chapter.

We have already given a one-dimensional version of this paradox in Figure 2.3. The single-photon source fires a photon from A in a straight line to the 50–50 beam splitter. According to the quantum theory, the photon leaves the beam splitter in a superposition of going to C and going to D. That is, half the energy of the photon heads to C. The other half travels to D. Since you cannot have half a photon, the photon wave collapses randomly, so all of the photon is detected either at C or D. Like the bubble, the detectors could be two lightyears apart. Then if D clicks the half of the photon that is about to hit C suddenly disappears from C and is transported instantly to D to combine with its other half, so D gets a full photon click. And then how does C know not to click if D clicks? It would seem that C and D are somehow conspiring to communicate with each other faster than the speed of light. That is the prediction of the quantum theory.

However, the experiment in Figure 2.3 has an Einstein-preferred hidden-variable interpretation. Suppose there is an invisible gremlin, undetectable by any means, which has a ping-pong paddle and who sits on the beam splitter. Further, suppose the photons are ping-pong balls. Each time a photon arrives at the beam splitter, the gremlin flips a coin. If it is heads, he whacks the photon into detector C. If it is tails, he whacks it into detector D. The outcome of this experiment is identical to that predicted by quantum theory but with no faster-than-light communication. However, the gremlin theory is not Newtonian mechanics or even classical statistical mechanics. It is a new theory specifically designed to be neither of those but to compete with quantum theory. Here the gremlin and his coin flips are the hidden variables. Since we posit that he and his coin are undetectable by any means, he is unknown. However, he is local. No spooky action at a distance between C and D is required to explain the gremlin theory. Einstein proposed that quantum theory might be replaced with something like gremlins with ping-pong paddles. It would then be a matter of taste if you believed in faster-than-light influences or undetectable demons. It turns out that – in a slightly different setup – only one of these two theories can agree with an experiment. This modified experiment is called the Einstein-Rosen-Podolsky paradox, and we will discuss it in the next chapter.

PARTICULARS OF POLARIZED PHOTONS

In this chapter, we have focused on the strange properties of a single photon. However, we have only considered the interplay of the wave and particle properties. As we know from chapter one, there is an additional property of the photon, and that is its polarization. Classically, the polarization is the direction of oscillation of the photon's electric field from the point of view that the photon is coming directly towards your eyeball. To make it easier to see it coming, we look at it through a gun-sight telescope, as in Figure 2.13. If the electric field is oscillating horizontally, we call it an H-polarized photon $|H\rangle$, and if it is oscillating vertically, we call it a V-polarized photon $|V\rangle$, as in the left of the figure. If, on the other hand, the electric field is oscillating at +45°, we call it a plus-polarized photon denoted by $|+\rangle$, and if the field is oscillating at −45° we call it a minus-polarized photon $|-\rangle$. (The photon can take on a host of different polarizations, but these suffice for our discussion.) What we want to demonstrate in this section is the unreality and uncertainty of a single photon's polarization. The photon polarization is a vector, and so we have the following vector-addition relations $|H\rangle + |V\rangle = |+\rangle$, $|H\rangle - |V\rangle = |-\rangle$, $|+\rangle + |-\rangle = |H\rangle$, and $|+\rangle - |-\rangle = |V\rangle$. It is here that the uncertainty and the unreality come to the fore. Remember Alice and her dog? One cannot say if the dog took the upper trail through the park or the lower path. The analogy

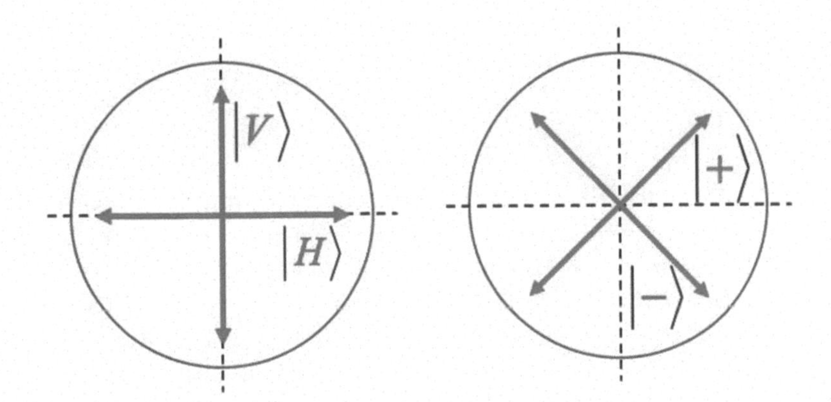

FIGURE 2.13 Photons can be polarized so that when they are coming straight toward you, their electric fields are oscillating either horizontally or vertically (left) and at ±45° (right).

here is that if $|H\rangle$ is the high trail and $|V\rangle$ is the lower trail, then the two states $|\pm\rangle$ are superpositions of the upper trail and lower trail. One cannot say which is the correct description; is the photon really $|+\rangle$ or is it really $|H\rangle+|V\rangle$. It is neither.

If we think instead of Schrödinger's cat and call $|H\rangle$ the live cat and $|V\rangle$ the dead cat, then the two states $|\pm\rangle$ are superpositions of the cat being both dead and alive at the same time. If I tell you that I have polarized the photon at +45° in the $|+\rangle$ state, you cannot know if the photon is really in that state or a superposition of $|H\rangle + |V\rangle = |+\rangle$. Only the experiment decides what state the photon is in, and the result depends on the measurement. In the original cat experiment, only two outcomes were allowed, dead *or* alive, but here we have two more consequences $|\pm\rangle$ that correspond to half dead *or half* alive. For large cats, we never see the latter, but for single photons, we see both, depending on the experiment. To view the effect, we need a polarization analyzer as shown in Figure 2.14.

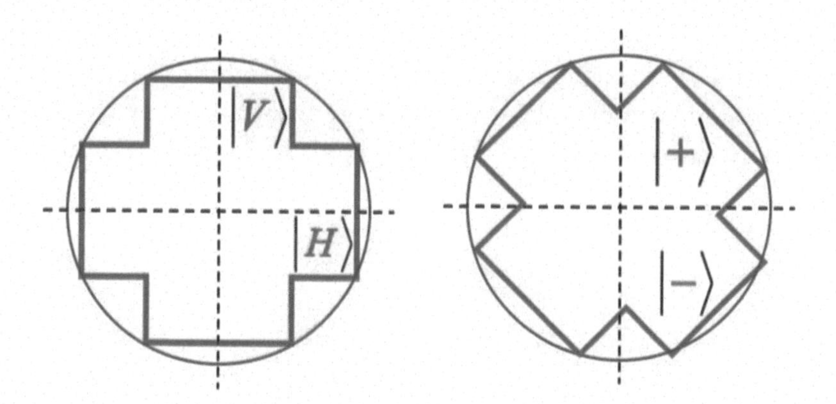

FIGURE 2.14 On the left, we show the detector set up in the horizontal-vertical (H–V) analyzing position and on the right, we show it set up in the ±45° analyzing position. If a horizontal photon hits the H–V analyzer, it passes through unscathed and is recorded by your eyeball as an H-photon. If a vertically polarized photon hits the H–V analyzer, it also moves unscathed and is registered as a V photon. Similarly, if ±45° photons hit the ±45° detector, they are also passed unscathed and recorded as being either plus or minus 45°. If these were the only possible outcomes, you might conclude that the polarization has an objective reality, and that we can detect its polarization with certainty. However, you would conclude wrongly.

Let us consider the four remaining options. Suppose an *H*-photon hits the ±45° analyzer on the right. That detector is set to pass only ±45° photons unscathed. What happens to an *H*-photon? Recall that we can write the *H*-photon as a cat state of ±45°-photons as $|H\rangle = |+\rangle + |-\rangle$, which is an equal superposition of a photon polarized at +45° and −45°. Since the ±45° detector is only able to pass either a +45°photon or a –45° photon unscathed, but not an *H*-photon, the *H*-photon does not pass unscathed, and the analyzer randomly collapses the cat into either a +45° photon or a −45° photon. *That's about as scathed as you can get.* Here quantum unreality and uncertainty again rear their ugly heads. Before the analyzer, the state of the *H*-photon is *unreal.* It could be an *H*-photon, or it could be a cat state of $|H\rangle = |+\rangle + |-\rangle$. Which is it? Reality is a myth. The experiment decides which it is. Uncertainty is here too in the mix, since which state you get, +45° or −45°, is entirely random. It is so random that physicists use this effect to make quantum-random-number generators. In quantum theory, it is impossible to say what the polarization of a single photon "really" is.

We can continue to concatenate these cats. Suppose the *H*-photon collapses to the +45° photon after passing through the first ±45° analyzer. If it is sent through a second ±45° analyzer, the +45° photon passes unscathed, leading you to believe it is really a +45° photon. However, if it is sent instead, secondly, through the *H*–*V* analyzer on the left, we have the same problem. The *H*–*V* analyzer is only able to pass *H* photons or *V* photons without harm. What does it now do to the *H*-photon that collapsed to a +45° photon? The collapsed +45° photon has no memory that it was once an *H* photon. However, we can write the +45° photon as a cat state of an *H* photon and a *V* photon as $|+\rangle = |H\rangle + |V\rangle$. Which is it really; a +45° photon or a cat state of an *H* photon and a *V* photon? It has no objective reality. The *H*–*V* analyzer collapses the +45° photon randomly either back to the *H*, from whence it came, with a 50% probability, or instead to the 90° rotated *V* photon also with a 25% probability.

To recap – if you send an *H* photon through a ±45° analyzer and then through an *H*–*V* analyzer, there is a 25% chance that what comes out is a *V* photon. (Because 50% of 50% is 25%.) How's that for unreality and uncertainty – what is that photon? Is it an *H*-photon or a *V*-photon. This question has no answer, and you'd best not ask it again. However, there is a somewhat complicated hidden-variable theory of this set up – that agrees with quantum theory. The theory involves replacing the photons with polarized ping-pong balls. Let us say the balls are spinning with their spin axes aligned with the *H, V,* or ±45° axes as in Newton's particle theory, the two analyzers are two coin-flipping gremlins with polarization rotators (spin rotators) instead of ping-pong paddles. Such a gremlin theory can reproduce the quantum theory – so far.

It is our goal in the next chapter to rule the gremlins out.

To conclude this chapter, we are back to the quote by Nobel Laureate Roy Glauber, "I cannot define a photon, but I know one when I see one." In this chapter, we have put a single photon through its many paces. I've tried to show you that quantum theory is not only weirder than you thought but weirder than you can think. However, without going into excruciating details about an approach that turns out to be wrong, all the experiments discussed in this chapter can be explained by dumping quantum theory and invoking unknown hidden-variable arguments. These arguments involve ever more convoluted Rube–Goldberg scenarios with multiple undetectable gremlins with cell phones, which are flipping coins and mucking about with our experiment.[46] Next, we'll show the gremlins cannot keep up with *two* photons.

---

46 I do go into agonizing detail about these gremlins in my previous book, *Schrödinger's Killer App*. Taylor and Francis, 2013. http://www.worldcat.org/oclc/1003668650. I do not have the heart to do it all over again here.

CHAPTER 3

It Takes Two to Tangle

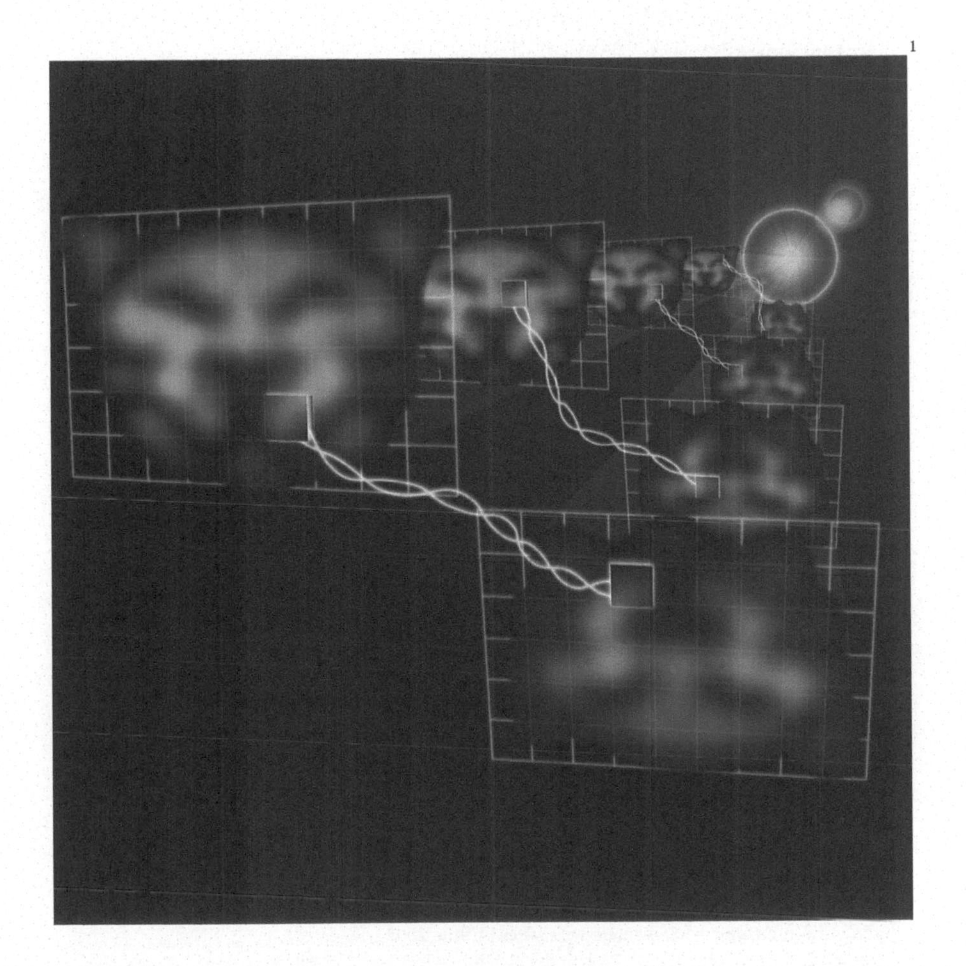

[1]

1 In this photomontage of actual quantum images, two laser beams coming from the bright glare in the distance transmit images of a cat-like face at two slightly different frequencies (represented by

> When two systems, of which we know the states by their respective representatives, enter into temporary physical interaction due to known forces between them, and when after a time of mutual influence, the systems separate again, then they can no longer be described in the same way as before, viz. by endowing each of them with a representative of its own. I would not call that one but rather the characteristic trait of quantum mechanics, the one that enforces its entire departure from classical lines of thought. By the interaction between the two representatives [the quantum states] have become entangled.[2]

SCHRÖDINGER'S CURSE

In my previous book, *Schrödinger's Killer App*, I go through the foundations of quantum mechanics in great detail, which probably negatively affected my sales. Here, for completeness, I will briefly review it again, and for all the gory details, you can procure my old book. I'd like to start with Schrödinger's Cat. I gave a lecture on this stuff some years ago to local high-school students and asked them, "Who here has heard of Schrödinger's Cat?" To my astonishment, they all raised their hands. I then inquired, "Where on Earth did you learn about that!" In unison, they replied, "*Big Bang Theory* – Duh!"[3] The foundations of quantum theory were now fodder for laughs on a favorite television comedy.

---

the orange and the purple colors). The braided lines indicate that the seemingly random changes or fluctuations that occur over time in any part of the orange image are strongly interconnected or entangled with the variations of the corresponding region in the purple image. Though false color has been added to the cats' faces, they are otherwise actual images obtained in the experiment. See also www.nist.gov/pml/div684/quantum_images.cfm. Credit: V. Boyer/JQI Disclaimer: Any mention of commercial products within NIST web pages is for information only; it does not imply recommendation or endorsement by NIST. Use of NIST Information: These World Wide Web pages are provided as a public service by the National Institute of Standards and Technology (NIST). With the exception of material marked as copyrighted, the information presented on these pages is considered public information and may be distributed or copied. Use of appropriate byline/photo/image credits is requested. Date: 11 June 2008.

2 Schrödinger, E. "Discussion of Probability Relations between Separated Systems." *Proceedings of the Cambridge Philosophical Society* 31 (1935): 555–563; 32 (1936): 446–451. See, https://doi.org/10.1017/S0305004100013554.

3 For a humorous take on the infamous cat, see, http://bigbangtheory.wikia.com/wiki/Schr%C3%B6dinger%27s_Cat.

As we see in Figure 3.1, we stuff the cat into a box with two halves. One half contains the cat. In the other half is a diabolical device containing a radioactive atom, a Geiger counter, a robotic arm, and a flask of cyanide. The atom has a half-life of a minute. That means that after a minute, there is a 50% chance it has decayed, emitting an alpha particle, which sets off the Geiger counter, activating the arm, smashing the flask of cyanide, and killing the cat. However, there is also a 50% chance that the atom has not decayed, in which case the cat is still alive.

Schrödinger pointed out that the quantum description of this theory was bizarre. According to quantum theory, after one minute, the atom is in a quantum superposition of decayed or not decayed. That is, the cat is in a quantum superposition of alive or not alive. Here are the two elements of quantum unreality and uncertainty at work. The state of the cat is unreal – it is neither dead nor alive until you open the box.[4]

FIGURE 3.1 Schrödinger's cat, as is familiar to anyone who watches the television show, "Big Bang Theory".[5]

4 It does not make sense to ask my niece, Christine Dowling, if her cat is dead or alive. She only owns a dog.

5 This file is made available under the Creative Commons CC0 1.0 Universal Public Domain Dedication. https://upload.wikimedia.org/wikipedia/commons/a/ae/Katze.jpg.

Similarly, the state of the atom is illusory, neither decayed nor not decayed. You may choose to open the box on the cat side and observe the state of the cat, or you may decide to open the atom side and measure the state of the atom. In either case, if you do this precisely at the one-minute point, the cat will randomly collapse into dead or alive with a 50–50 probability, and the atom will collapse into decayed or not decayed, with the same chance. This randomness is quantum uncertainty at work. Which result you get is entirely random and unpredictable by any means.

Now we add the final element of Einstein's discontent. Quantum theory predicts that these results are independent of distance. We can ship the cat to the star system Alpha Centauri and send the atom to Beta Pictoris, which are 62 lightyears apart. If we put them both in ships moving at nearly the speed of light, they will arrive at their destinations within a minute in their reference frame, due to the relativistic time-dilation effect of the Einstein twin paradox.

Now suppose Bob, sitting on Beta Pictoris, opens his half of the box, and collapses the atom at random to decayed. His action will instantaneously collapse the cat to dead – 62 lightyears away. This setup is quantum nonlocality at work – measurements here immediately affect the measurements made lightyears apart.

Einstein did not like *that* one bit, as that prediction seemed at odds with his theory of relativity. Schrödinger called this weird correlation between the atom and the cat "quantum entanglement" in his cat paper. He was trying to imply by this paradox that quantum theory could not possibly be right because this prediction of collapse over lightyears was nonsense. Well, Schrödinger was wrong, and Einstein was wrong, and quantum theory is correct. Let's see how.

EINSTEIN'S BANE

The Schrödinger's cat paper appeared in November of 1935.[6] It was an attempt to explain an earlier article by Einstein and his two collaborators, Boris Podolsky and Nathan Rosen, which had appeared in May of

6 Here is the more detailed description of the fell feline https://commons.wikimedia.org/wiki/File:Katze.jpg.

1935 with the title, "Can Quantum-Mechanical Description of Physical Reality be Considered Complete?" To this hypothetical question, Einstein and his pals replied with a resounding *no*. In this paper, now called the EPR paper (after the initials of the three authors' last names),

EINSTEIN ATTACKS QUANTUM THEORY

Scientist and Two Colleagues Find It Is Not 'Complete' Even Though 'Correct.'

SEE FULLER ONE POSSIBLE

Believe a Whole Description of 'the Physical Reality' Can Be Provided Eventually.

FIGURE 3.2 Clipping from the New York Times from 04 May 1935 upon the publication of the EPR paper.[7] One of the authors, Boris Podolsky, leaked a preprint to the New York Times in advance of its release, and Einstein was so mad that he never spoke to Podolsky again. The EPR paper has over 10,000 citations, more than all of Einstein's other papers combined.

7 The clipping from the New York Times comes from here and is not subject to U.S. copyright, https://commons.wikimedia.org/wiki/File:NYT_May_4,_1935.jpg.

Einstein launched an attack on the quantum theory. He targeted the three features he hated most, quantum unreality, quantum uncertainty, and quantum nonlocality. The paper was a clarion call to all the quantum-doubting Thomases and made a splash in the *New York Times*. See Figure 3.2.

Einstein pointed out that all three of these horrible features exist in quantum theory, and all three of them were unacceptable – at least to him. He then laid out a program to replace quantum mechanics with a probabilistic approach that was simultaneously real, certain, and local. How could he do this? His competing theory we now call a hidden-variable theory.[8] Let's think about our photons and their polarizations.

When I send a +45°-polarized photon through an *H*–*V* polarization analyzer, the state of the photon is *unreal* before I make a measurement then it collapses to *H* or *V* with a 50–50 probability, and the outcome is unpredictable by any means. Let's compare this to a classical coin toss. Suppose I flip a nickel coin into the air and slap it down onto the back of my hand and yell, "Heads or tails?" Well, you can choose heads, or you can choose tails. You know it is one or the other, but you don't know which. In the quantum version, it is *neither* one *nor* the other, and you can *never* tell until you look.

But a coin is a classical object like a cat, and so we can describe it with Newton's laws of mechanics. If you knew the precise way that I flipped the coin and knew the angle, acceleration, forces, velocity, air friction, and details of the shape of my hand, you could use a supercomputer to predict the coin's final position – heads or tails. But in practice, nobody has access to all that information, so we sweep it under the rug and are happy with *either* heads *or* tails. The point is if that you did know all that information, then the coin toss would be classically real – the coin is either heads or tails, but you don't know which – and it is classically certain. You could predict which if you knew all that data. Such a theory of the traditional coin toss is called a hidden-variable approach. Classically, all forces and velocities and accelerations and the grittiness of my hand are in principle knowable to you but, in practice, unknown to you. Therefore,

---

8 A detailed description of the EPR paradox is here, https://en.wikipedia.org/wiki/EPR_paradox.

you average over them. This description is a classical statistical theory of the coin toss. To compare, in quantum theory, there are no hidden variables and the outcome of the coin toss in unknowable to you – even in principle.

Einstein posited that perhaps quantum mechanics could be replaced with a classical hidden-variable theory. All the weirdness goes out the window, and quantum theory becomes a branch of probability theory. Einstein then also pitched the notion that a classical hidden-variable theory could take care of the weird quantum nonlocality. Then quantum mechanics would be a sensible theory that people would stop arguing about. To see how the nonlocality is removed, we have to come up with a hidden-variable theory of entanglement. To do this – we need more nickels!

At the beginning of next month, the Department of Entanglement sends to you and me two 12-count egg cartons labeled "entangled nickels inside". When we open our respective containers, we see that each of the 12-egg pockets holds a single nickel coin, and that they have numbered the pockets in order, 1–12. According to a pre-arranged protocol, for the entire month, we flip the coins in order of their numbering and write down whether we got heads or tails for that flip. We do not coordinate our actions. The only requirement is that we flip them in order. I might decide to flip one every other day or so, and you may choose to flip them all on the last day. Then at midnight on the last day of the month, we call each other up and compare the tally. Our records are the same. Whenever you got heads, I got heads, and whenever you got tails, I got tails. If we repeat this experiment many times, we always get this miraculous result that our coin flips match. Now, what would *you* say is going on?

The first thing you might guess is that it is some magic trick – a sleight of hand! – which the DoE is playing on us. Just like any magic trick, it looks magic until the magician lets you in on the secret. Then all is revealed. It is that magician's secret, which is the hidden variable. Even though we do not know the trick, we should not be astonished, because we know there is always a trick. After all, there is no such thing as magic. What could the trick be? Well, maybe the DoE put little gyroscopes in the coins so that, apparently random to us, some pairs were programmed to produce heads and other couples to produce tails.

Maybe there are small invisible alien beings shipped with the coins, that only the DoE is aware of, which they have trained to use their tractor beams to make some pairs go heads and other pairs go tails. It does not matter, like in the magic show, we know it is a trick, but we don't know the details. That was Einstein's bane, he thought the coins are all prewired by hidden variables that we either do not know - or cannot know - but nothing magic is going on!

However, this outcome of our experiment is also precisely what quantum theory predicts if the coins are entangled, like the cat and the atom. However, the quantum version of this double-coin-tossing experiment is much stranger than even invisible microscopic aliens with tractor beams. In the quantum version, we have prepared the coins in the entangled state $|H\rangle_{\text{you}}|H\rangle_{\text{me}} + |T\rangle_{\text{you}}|T\rangle_{\text{me}}$. Here, you are the cat, and I am the atom, and H is analogous to cat alive, and atom not decayed, while T is that of cat dead and atom decayed. Of course, they now stand for the nickel coins' turning up heads or tails, since the People for the Ethical Treatment of Animals frown on such experiments with real cats. The weird brackets indicate quantum states as before.

Now according to quantum theory, the coins are *neither* heads *nor* tails until we flip them. Then, when we do, if my coin collapses to heads, yours simultaneously collapses to heads, with a 50% probability that is unpredictable by any means. And this happens if you are in Alpha Centauri and I am in Beta Pictoris. It takes light about 62 years to go between them, but the collapse occurs instantaneously. If you flip first, then you collapse my coin, and again there is a 50–50 chance you will get heads or tails, but whatever you get, my coin collapses to the same thing. Then at the end of the month, we see the same results as the hidden-variable interpretation.

At play here are the three weirdnesses of quantum theory. *Unreality* - the coin is neither heads nor tails until flipped. *Uncertainty* - the coin collapses to heads or tails with a 50–50 probability that is unpredictable by any means. *Nonlocality* - the two coins always give the same results, either heads–heads or tails–tails, even though they are a gazillion miles apart.

It *could* be that the hidden variable magic trick could account for this experiment just as well as quantum theory. Einstein's dream was to replace the quantum theory with invisible aliens toting tractor beams. You have to realize how desperate he was to prefer the latter to the former. But invisible aliens with tractor beams can be described by a theory that is real, certain, and local, and that's all that mattered. Einstein hoped that quantum mechanics was a magic trick played upon us by a non-dice-tossing God who had decided to hide the trickery from us, but, that in *her* benevolence, had made the fraud decidedly classical.

For the above experiment, we can show that the outcome of the hidden-variable theory with invisible aliens gives the same predictions as the quantum theory. The hope was that if that were true for all experiments, then it was just a matter of taste which method you decided to use to explain quantum phenomena – invisible aliens or quantum theory. At this point, it was all a stuffy debate for the philosophers.

Many famous quantum physicists, such as the renowned Danish physicist Niels Bohr, recoiled at Einstein's program of invisible aliens, but they were not able to prove that it was wrong. And so, the hidden-variable theory hung around our necks for decades like the moldering carcass of a decaying albatross. But then, apparently, one of the great minds of quantum theory – John von Neumann – came to our rescue – or so we thought.

VON NEUMANN'S BLUNDER

John von Neumann was a Hungarian physicist, with a keen mathematical mind, who immigrated to the U.S. just ahead of the Nazi takeover and helped develop the atomic bomb. (See Figure 3.3.) Here in America, he developed a taste for big fancy suits, big American cars, and big cigars.[9] He was a founder of many fields of mathematics, physics, and computer science. In quantum physics, he is best known for his 1932 book, *The Mathematical Foundations of Quantum Mechanics*. Even though written three years *before* Einstein's paper, von Neumann was prescient enough to preemptively strike back – forward? – and provide a mathematical proof that no hidden-variable theory could ever agree with all the predictions of quantum theory. That is, he claimed they were two different theories and not two different philosophies. Implicit in his result was the idea that they should give different results in the lab, but he did not spell out how. Von

9 Hargittai, István. *The Martians of Science: Five Physicists Who Changed the Twentieth Century*. Kindle Locations 1451–1453. www.worldcat.org/oclc/213468654.

Neumann was brilliant, so most of the community simply quoted this result to put the hidden-variable argument to rest. "The phrase 'von Neumann has shown …'" entered the quantum physicist's lexicon as a debate ender.[10]

However, Grete Hermann, a brilliant young German mathematician and philosopher, read the proof and found a mistake. She reported her finding to another founder of quantum theory, German physicist Werner Heisenberg (of uncertainty principle fame).[11] She published this critique of von Neumann's proof in the same fateful year as Einstein and Schrödinger, 1935. Alas, perhaps because it was too early, or because she was a woman,

FIGURE 3.3 While I would like to say that this is John von Neumann's mug shot from his arrest for reckless driving, it is his ID badge photo from when he was working at Los Alamos on the atomic bomb.[12]

10 Gilder, Louisa. *The Age of Entanglement* (Kindle Locations 2612–2613). Kindle Edition. Knopf Doubleday Publishing Group, 2009.

11 The life and times of Grete Hermann, https://en.wikipedia.org/wiki/Grete_Hermann.

12 Unless otherwise indicated, this information has been authored by an employee or employees of the Los Alamos National Security, LLC (LANS), operator of the Los Alamos National Laboratory under Contract No. DE-AC52-06NA25396 with the U.S. Department of Energy. The U.S. Government has rights to use, reproduce, and distribute this information. The public may copy and use this information without charge, provided that this Notice and any statement of authorship are reproduced on all copies. Neither the Government nor LANS makes any warranty, express or implied, or assumes any liability or responsibility for the use of this information. https://commons.wikimedia.org/wiki/File:John_von_Neumann_ID_badge.png.

the physics community universally ignored her result. Her critique remained forgotten for 30 years until John Bell, a Northern-Irish physicist, independently rediscovered it. In the meantime, the community continued to refer to von Neumann's "proof" as a debate ender.

BOHM'S BOMBSHELL

David Bohm was a tragic figure and an American physicist. He had joined the communist party as a youth, and then became caught up in the McCarthy red scare of the early 1950s and found that he could no longer find a job. Bohm fled to Brazil in 1951, where he found a professor position at the University of São Paulo. Upon his arrival, the U.S. Government promptly canceled his U.S. passport. This act of pettiness frightened him to the point that he became paranoid, thought he was being followed everywhere, and found he could not eat any of the local food without becoming ill.

While in Brazil, one of his most important papers appeared, in which he constructed a hidden-variable theory that agreed with the predictions of quantum theory. He did what von Neumann said was impossible. His theory gobsmacked some in the physics community, but most ignored him, as they had ignored Grete Hermann. In 1955, he relocated to Israel, where he found a better job, a better climate, and edible food.[13]

Bohm had reinvented something that had been discovered many years before by another founder of quantum theory, the French physicist Louis De Broglie, who called the approach the pilot-wave theory. The de Broglie–Bohm theory is just as crazy as either quantum theory or invisible aliens with tractor beams. To explain the two-slit diffraction experiment, described in the first chapter, Bohm fills the entire universe with an invisible force field that is undetectable by any means. The field pushes the photons around in such a way that that it recovers the predictions of quantum theory. Their theory was real – the photons had a reality like ping-pong balls. Their method was certain – you could predict exactly where they would go. (See Figure 3.4.)

It all sounds good so far. But their approach had a flaw, the one that Einstein *most* loathed. The theory was *nonlocal* – if you closed one of

---

13 More on the sad story of Bohm, https://en.wikipedia.org/wiki/David_Bohm.

the slits on Alpha Centauri, the invisible force field would instantaneously shift around on Beta Pictoris, just in time to guide the photons to the right place. Many people like this theory. They are called Bohmians. I think the Bohm theory is just as weird as quantum theory, and quantum theory is more natural for me to work with. I don't dislike the Bohm theory – I simply have no use for it.[14] But the

FIGURE 3.4 The de Broglie–Bohm pilot wave interpretation of the two-slit diffraction experiment. The blue lines represent the nonlocal potential or pilot wave that extends to infinity. The pilot wave is the hidden variable. The yellow lines are the deterministic paths the photons take. The photons do not interfere but look like they do. Blocking one of the slits on the left instantaneously changes the pilot wave potential's shape arbitrarily far away to the right. Hence the theory is certain, it is real, but it is nonlocal.[15]

14 The origin of the Bohm hidden-variable theory, https://en.wikipedia.org/wiki/De_Broglie%E2%80%93Bohm_theory.

15 This file is licensed under the Creative Commons Attribution-Share Alike 4.0 International license, https://commons.wikimedia.org/wiki/File:3_trajectories_guided_by_the_wave_function.png.

approach made an impact in certain quarters, for those who were paying attention, and who had never heard of Grete Hermann – the success of the Bohm theory meant that von Neumann's unassailable impossibility proof was *wrong*.

BELL GETS TESTY

John Stuart Bell was a theoretical physicist whose day job was designing those big atom smashers at CERN – that's the job that paid his bills.[16] But at night and on weekends, he would think about the foundations of quantum theory and puzzle over things like the EPR paper, Bohm's approach, and von Neumann's no-go theorem. I don't like no-go proofs as they often depend on hidden assumptions and can shut down a whole field of research with one rebuttal, "von Neumann showed it could not be done." Like Grete Hermann years before, but this time motivated by Bohm's paper, Bell realized that the Bohm theory implied that the von Neumann proof had to be wrong. He dug through the proof and found the same mistake Hermann found – von Neumann had made the implicit assumption that the hidden-variable theory was local – no spooky action at a distance.

Bell then realized that the Bohm model was a *nonlocal* hidden-variable theory – just as quantum theory was! Nonlocality was the most reviled of Einstein's three bugaboos – it was not on his program to replace quantum theory with a *nonlocal* hidden-variable theory. Einstein explicitly declared that the hidden variable argument had to be real, certain, and *local*. Bohm's approach had met the first two requirements but not the last. Hence, Bell set out to redo the von Neumann proof and make explicit Einstein's locality requirement. Bell's work resulted in his 1964 paper, "On the Einstein Podolsky Paradox." Here, he provided a simple proof that no *local* hidden-variable theory could reproduce all the predictions of quantum theory. To do so, the unknown variable had to contain at least two of Einstein's three no-nos. Either it had to be uncertain and unreal, or it had to be uncertain and nonlocal. Of course, it could be all three – uncertain, unreal, and nonlocal – just as quantum mechanics is.

---

16 For whom the toll Bells, https://en.wikipedia.org/wiki/John_Stewart_Bell.

The proof is now called Bell's theorem, and it too was almost universally ignored. By the 1960s, quantum physics had entered its shut-up-and-calculate phase, where even thinking about the foundations of quantum mechanics could be career limiting. In 1984, when I started my Ph.D. research into the foundations of quantum theory, a member of my dissertation committee told me that it was crackpot stuff and that I would never get a job.[17] In Bell's paper, he made the astonishing claim that you could test the two theories against each other in the lab. He explicitly spelled out how. We are no longer dealing with metaphysics – we are dealing with physics!

In Einstein's original proposal, the idea was that the local hidden-variable theory should always agree with the quantum theory, and we could use either one as a matter of taste. Bell showed that these were not two equally good interpretations of one theory, but rather two completely different theories, which made different predictions in the real world. To quote the theoretical physicist Hans Bethe, "It is the job of every theoretical physicist to produce a number that can be compared with an experiment – and if you're not doing that – then you are not doing your job!"[18] Bell had done his job. He produced a number that we could compare to an experiment. This proposed setup is now called the Bell test.

We show the Bell-test experiment in Figure 3.5. I blathered on endlessly about this test in my previous book. However, since this test is an essential component of the quantum internet – perhaps the most crucial element – I'll briefly review it again here. The source Stevie at S emits a pair of entangled photons in the state $|H\rangle_{\text{Alice}}|V\rangle_{\text{Bob}} - |V\rangle_{\text{Alice}}|H\rangle_{\text{Bob}}$, where Alice is A on the left in Alpha Centauri, and Bob is B on the right

---

17 As a graduate student, I had taken all my quantum courses from Prof. K.T.M. I was his best student. After I passed the qualifying exams in January of 1984, he expected me to knock on his door I suppose and declare that I wanted to work with him on superstring theory. Instead, I signed up with his archenemy, Prof. A. O. Barut, to work on quantum foundations. When K.T. M. found out about this, he tracked me down at a student-faculty party in the lounge on top of the physics tower, and after several glasses of red wine, he grabbed me by the arm with one hand, and scolded me with the other hand, "Dowling! I hear you have signed up with Barut to do quantum foundations! That is crackpot stuff – I tell you – you will never get a job!" I held my ground, and in July of 1994, I emailed him a three-line message, "KTM! I got a job! Dowling!" He did not reply.

18 Something he said in a lecture I attended at Caltech circa 2000, and you can find more about him here https://en.wikipedia.org/wiki/Hans_Bethe.

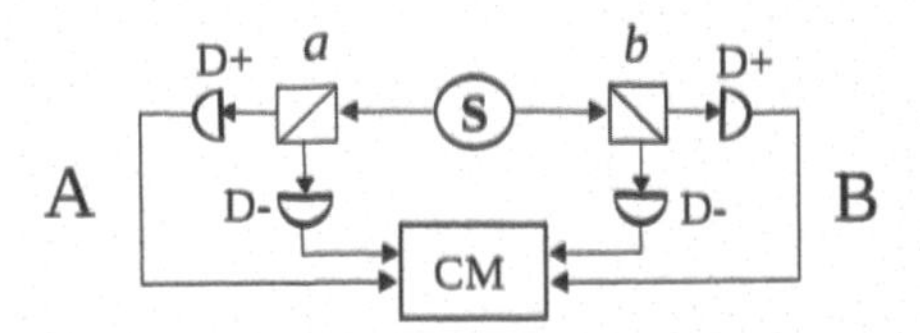

FIGURE 3.5 A Bell-testing device for entangled pairs of polarized photons. The S is the entangled photon source sending a couple of photons to Alice and Bob labeled A and B. Alice's polarization-analyzer angle is at a and Bob's is at b. D± are photodetectors. The box labeled CM is a coincidence monitor that takes Alice and Bob's measurements and polarization settings and plugs it into the Bell-test formula. If the machine spits out a number less than two, Einstein's hidden-variable theory is right. If it spits out an amount bigger than two, then quantum mechanics is correct. In practice, at least in the early experiments, the numbers were just barely bigger than two, with a large margin of error, so you had to average over many thousands of photon-pairs to be sure it was bigger than two.[19]

in Beta Pictoris.[20] This experiment is similar to our previous analysis with the nickel coins in the egg crate, except we have replaced the coins with polarized photons. Alice and Bob each have an adjustable polarization analyzer, as discussed in chapter two. To reproduce the entangled-coin experiment, Alice and Bob both set their analyzers to the *H*–*V* position. This state is a bit different than that used with the coins.

Here, Alice and Bob's measurements are anti-correlated. That means when Alice gets an *H* photon, then Bob always receives a *V* and vice versa. Let's send 12 pairs of photons like the 12 nickels. Alice and Bob order them by their time of arrival. The first photon to arrive is photon one, etc. Alice and Bob can store the photons in optical delay lines, keeping them ordered, and then follow the same protocol as with the nickels. Within a month's period, they choose to measure their photons' polarizations at random times. At the end of the month, they compare notes and find that whenever

19 Figure taken from https://commons.wikimedia.org/wiki/File:Two_channel_bell_test.svg. Permission is granted to copy, distribute and/or modify this document under the terms of the GNU Free Documentation License, Version 1.2 or any later version published by the Free Software Foundation; with no Invariant Sections, no Front-Cover Texts, and no Back-Cover Texts.

20 A shout-out to Stevie Turkington – my biggest fan!

Alice got a V, then Bob got an H, and vice versa. This result is as strange as the coin experiment, and of course, you would suspect the source is somehow fiddling with the photons. For concreteness, we'll put the source in the middle between the star systems, discussed above. It takes a photon 31 lightyears to reach Alice and 31 lightyears to get to Bob.

Again, we expect it is a trick. We may not know what the sleight-of-hand is, but we feel sure that somebody on the source, Stevie, is setting us up by doing something to the photons, or sending undetectable aliens along with the photons. Stevie programs the aliens in advance to fool around with the photons, so when Alice gets an H, Bob always receives a V, and vice versa. The point of the Bell test is to rule out any such trickery. To do this, we need to add a few more ingredients until we reach a point that the predictions of quantum theory and hidden-variable theory diverge. For the current setup, the quantum interpretation is that measurement causes the entangled photons to collapse randomly into H–V or V–H with a 50–50. The hidden-variable theory would then consist of undetectable aliens that are launched from the source and preprogrammed to deflect the photons to get the same result – no way to distinguish these two theories – yet.

The particular entangled quantum state $|H\rangle_{\text{Alice}}|V\rangle_{\text{Bob}} - |V\rangle_{\text{Alice}}|H\rangle_{\text{Bob}}$ is called an EPR state or a Bell state, and it has the peculiar property that it looks the same in any polarization basis. Using the results from the last chapter, we can show that $|H\rangle_{\text{Alice}}|V\rangle_{\text{Bob}} - |V\rangle_{\text{Alice}}|H\rangle_{\text{Bob}} = |+\rangle_{\text{Alice}}|-\rangle_{\text{Bob}} - |-\rangle_{\text{Alice}}|+\rangle_{\text{Bob}}$. Coins can't do this! So now Alice and Bob redo the experiment. The source sends out 12 of precisely the same entangled states, but now Alice and Bob always measure in the ±45° basis instead of the H–V basis. After a month, they compare notes, and when Alice got a +45° result, then Bob always got a – 45° result, and vice versa. There is no right analog here with coins, because photons are unreal. This result takes us back to chapter two – the source does not determine what state the coin is in – it is the measurement that does. In this way, we can rule out hidden variable trickery and reveal the true quantum magic.

The next step is to make it impossible for the source to send preprogrammed aliens that will conspire by mobile phone to cause the hidden-variable theory to mimic quantum theory. To do this, Alice and

Bob now agree that, a few seconds before each photon's arrival, they will use a quantum-random-number generator to randomly choose the *H*–*V* analyzer or the ±45° analyzer and even weirder angles in between. At the end of the month, they now only compare the measurements, about six on average, where they accidentally chose the same analyzer. For those events only, whenever Alice gets an *H*, Bob receives a *V*, and vice versa, or whenever Alice receives a +45° result, Bob always gets a −45, and vice versa. Now Stevie at the source cannot know in advance which analyzer Alice and Bob will use, because he is 31 lightyears away when they make their choice. The preprogrammed aliens would have to communicate faster than the speed of light – indeed, infinitely fast – to continue to fake out Alice and Bob. But things moving instantaneously fast are the hallmark of the Einstein-reviled nonlocality. The hidden-variable theory, like the theory of Bohm, would have to be nonlocal to agree with quantum theory. But it is precisely that *non*-locality we are trying to rule out.

The test is then straightforward. Alice and Bob randomly choose different angles for their polarization measurements. A typical scenario is for Alice to decide to start in the 0°/45° basis and then randomly rotate this around in chunks of 22.5°. Bob does the same. Note that we now allow the analyzer to have nonperpendicular angles (such as *H*–*V* and ±45° are), which is critical to finally rule out any local hidden-variable theory. This time they don't throw anything away. They make a record of what they measured and what the analyzer was set to. At the end of the month, they call each other up and take their combined data and stick it into a formula from Bell's paper. The formula spits out a single number. If that number is bigger than two, then the quantum theory is ruled in. If that number is less than two, then the hidden-variable theory is ruled in. A few experiments were carried out in the 1970s, but the first that had to be reckoned with was the 1972 experiment of Stuart Freedman and John Clauser. I will discuss it briefly below. For more excruciating details, read my previous book, or get thee to the Internet.[21]

21 A somewhat incomplete history of the Bell test experiments may be found here, https://en.wikipedia.org/wiki/Bell_test_experiments.

BERTLMANN'S SOCKS

John Bell had an eccentric colleague at CERN, the Austrian physicist Gerhard Bertlmann. Bertlmann had this odd habit of always wearing one red sock and one green sock, as shown in Figure 3.6. Using the socks as a prop, in 1981, Bell wrote a now-famous paper, "Bertlmann's Socks and the Nature of Reality."[22] If all my coin-flipping explanations of the Bell test for hidden variables versus quantum theory left you in a vegetative state, here, I will try paraphrasing Bell's own words again.

As you can see, Bertlmann always wears this peculiar sock combination. Suppose Bertlmann is coming around the corner in the hallway ahead of you. The first thing you will see is either his right or left foot – let's say it's his left (that we can see in the photo has a red sock). You can then immediately conclude that when his right foot appears into view, the sock will be green. Half the time, Bertlmann chooses left-foot red and right-foot green while dressing himself in the morning, and the other half the time, he picks the opposite option. Why are your observations about Bertlmann's socks different than the predictions of quantum theory? It is because, once he makes his choice, the colors are fixed. Bertlmann himself is the hidden variable theory. If you hid a spy camera in his dressing room, you would know what color combination to expect at work.

CLAUSER'S NO-BELL PRIZE

In 1972, John Clauser was a postdoctoral researcher at the University of Berkeley in California, and Stuart Freedman was a graduate student there. (See Figure 3.7.) Clauser had read Bell's paper and decided to put

22 In November of 2017, I was visiting the University of Vienna, where Bertlmann is now a professor. I had never met this curious gentleman, but one day while chatting with Anton Zeilinger in his office, an otherwise well-dressed fellow rushed in to ask him a question. Zeilinger waived off the question and then politely introduced him to me, "Prof. Dowling, I'd like to you to meet Prof. Bertlmann." I gasped and exclaimed, "Not *the* Bertlmann!" Without a word, Bertlmann yanked up his trousers and proudly displayed his infamous non-matching socks. I immediately snapped this photo of his feet. He then inquired, "Would you also like a photo of my face?" I replied, "Nah. Nobody knows what *you* look like, but everybody knows what your *socks* look like!" I tried to get Zeilinger to take a photo of us both from head to toe, displaying our socks, but the world-famous experimenter could not work the camera on my phone and then – in desperation – threw both of us out of his office.

FIGURE 3.6 John Bell had a slightly eccentric colleague by the name of Bertlmann, who always wore one red sock and one green sock, as shown.[23] Bertlmann became famous when Bell wrote a paper featuring him and his odd choice of footwear.

its predictions to the test. The setup of his experiment is precisely like that in Figure 3.5. The entangled photon source produces many entangled-photon pairs of the form $|H\rangle_{\text{Alice}}|V\rangle_{\text{Bob}} - |V\rangle_{\text{Alice}}|H\rangle_{\text{Bob}}$, of which about 100 pairs are counted per second by Alice's detectors (A) and by Bob's (B). The polarization analyzers *a* and *b* can be set to any pair of polarization angles, not just ±45° or *H*–*V*. In this first experiment, Clauser did not have the delayed choice aspect of randomly

23 "Bertlmann's Socks and the Nature of Reality," by John S. Bell, J. Phys. (Paris) 42, C2-41 (1981); "Is the moon there when nobody looks? Reality and the quantum theory," PHYSICS TODAY/ APRIL 1985 PAG. 38–47. A link to the former is here http://cdsweb.cern.ch/record/142461/files/ 198009299.pdf. A link to the latter is here. https://cp3.irmp.ucl.ac.be/~maltoni/PHY1222/mer min_moon.pdf.

FIGURE 3.7 John Clauser (left) holding forth in 2010 at a conference coffee break with American physicist, Michael Nauenberg (right) listening closely.[24] Clauser always wore the same brand of polo shirt – I forget which brand – but it was undoubtedly not Polo. They were always solid in color and had a pocket for all his pens. In his honor, I, too, now always wear off-brand solid-colored polo shirts with a pocket.

changing the polarization settings after the photons left the source. What you need to know is that Clauser expected to win the Nobel Prize by ruling quantum theory out and Einstein in. He kept taking more and more data, and the number that the Bell-tester algorithm kept spitting out was always statistically significantly larger than the magic number of two. Clauser had to believe his data – Einstein was wrong. No Nobel Prize for Clauser – yet! – but in 2010 he won the Wolf Prize in Physics for this work.

In 1976, he repeated this experiment with even better statistics and ruled Einstein's hidden-variable theory out with more certainty. By this time, the experimenters realized that the maximum violation occurred

24 Photo from https://commons.wikimedia.org/wiki/File:John_Clauser_conversing_with_Mike_Nauenberg.jpg. This work is free and may be used by anyone for any purpose. Author: Nick Herbert.

when the Bell test spits out $2\sqrt{2} = 2.83$, and that happened for the oddball settings of the analyzers when Alice places hers at 0° and 45° while Bob places his at 22.5° and 67.5°. In all modern experiments, instead of rotating the polarizers around, you typically sit at these angles. Based on the quantum random number generator at A, Alice randomly chooses either the 0°/45° basis or the 22.5°/67.5°, and Bob does the same. They keep the measurements only from the event where they chose the opposite basis. They plug that data into the Bell tester, and if everything is perfect, it spits out 2.83. However, in the early experiments, almost nothing was perfect. So the number it expectorated was much closer to 2.00 than it was to 2.83. However, as the experiments improved, the number moved inexorably upward to the 2.83 limit. Current tests show that quantum theory is right and Einstein's theory is wrong at a confidence level of 99.999 … %, where there are hundreds of nines in there.

However, another prediction is that quantum theory says these weird nonlocal effects do not drop off with distance. In the 1970s, Clauser carried out some of the first experiments with Alice and Bob separated by about 10 m. In the most recent 2017 test - carried out with the satellite Mozi by the group of Chinese physicist Jian-Wei Pan - the Bell test ruled quantum in and Einstein out with Alice and Bob separated by 12,000 km. There is no sign that these effects disappear or degrade with distance.

There are four two-photon entangled states, called the Bell states, and they are $|\Phi^{\pm}\rangle_{AB} = |H\rangle_A|H\rangle_B \pm |V\rangle_A|V\rangle_B$ and $|\Psi^{\pm}\rangle_{AB} = |H\rangle_A|V\rangle_B \pm |V\rangle_A|H\rangle_B$. Given any one of them, we can manipulate it into any other using polarization rotators. In this way, the Bell states are all just one kind of state. We now have very bright sources that can produce any one of these at the rate of about a million pairs per second, compared to Clauser's rate of about 100 pairs per second. This bright source is mounted on the Chinese satellite. As we shall show, the quantum internet will be composed of nodes that do nothing but spew out these Bell states to other members on the network. Once Alice and Bob share a large number of Bell states, they can use them to do quantum teleportation, distributed quantum computing, build a quantum repeater, and carry out quantum cryptographic key distribution. All of these protocols allow us to move quantum information around on the network securely. One protocol we'll discuss is the entanglement-based cryptography protocol developed in 1991 by Polish physicist Artur Ekert. His scheme uses the Bell test and entangled photon pairs to establish an unbreakable cryptographic key. When Ekert showed this idea to Bell, Bell exclaimed, "Are you telling me that this could be of practical use!?"

Little did Bell know what quantum technologies he had unleashed.[25]

CLOSING THE LOOPHOLES

As the experimentalists made better and better Bell testers, the theorists made ever more improbable hidden variable theories. These gave rise to loopholes because while these theories were unlikely, they were not impossible.[26] The two most important are the detector loophole and the locality loophole. In the early days, the detectors were terribly inefficient, catching less than 50% of the photons striking them. This flaw allowed for a hidden-variable theory that consists of two undetectable gremlins, sitting on the detectors, skewing your data to make you think quantum theory is correct. The mischievous demons make up the local-hidden variables. I doubt that Einstein would accept invisible pixies, but the known laws of classical physics cannot rule them out. The point is to jettison them out of the detectors by exploiting the laws of *quantum* physics.[27] In practice, if the detectors' efficiency is less than 83%, the Bell tester always spits out a number less than two, making you think quantum theory is wrong. To close this loophole, we had to wait some years for the detectors to improve to better than this 83% efficiency. David Wineland's group at the National Institute of Standards and Technology (NIST) first closed this loophole in 2001.

Now the locality loophole is a different can of gremlins. The problem here is with the delayed-choice part of the Bell test. Alice and Bob must

---

25 See, Gilder, Louisa. *The Age of Entanglement.* Kindle Location 5942. Knopf Doubleday Publishing Group, 2008.

26 Closing the loopholes. https://en.wikipedia.org/wiki/Loopholes_in_Bell_test_experiments#Detection_efficiency,_or_fair_sampling.

27 For an exhausting explanation of the detector and locality loopholes, see the section, Fairies, Gremlins, and Magic Dice, starting on page 19 in chapter one of my book, *Schrödinger's Killer App: Race to Build the World's First Quantum Computer.* Taylor and Francis CRC Press, 2013. www.worldcat.org/oclc/1003668650.

choose their detector settings randomly and quickly and far enough apart to rule out gremlins on the source. Delayed choice also keeps the detectors from conspiring with each other to trick us into thinking quantum theory is wrong. Here, the gremlins with cell phones are the hidden variables. To rule the gremlins out, Stevie the source, Alice's detector, and Bob's detector all three must be so far away from each other that, for the demons to fool us, they must be able to communicate faster than the speed of light. While quantum theory is nonlocal, it does not permit the transmission of information more quickly than Einstein's speed limit. Since mobile phone signals travel at that speed or less, we can rule the gremlins out. The Austrian group of Anton Zeilinger, using a separation between Alice, Bob, and Stevie of over several hundred meters, first conclusively closed the locality loophole in 1998. It was last closed over a distance of 1,200 km by Jian-Wei Pan's group in China using an entangled photon source on their quantum communications satellite, and Alice and Bob on two ground stations, separated by that distance.

However, there was still a problem. We had to close the detector loophole and the locality loophole simultaneously to rule out a particular class of uniquely evil gremlins. In my 2013 book, I predicted that it would happen in the next few years, and it would be done by the Austrian group or by the U.S. group at NIST. I got both right but missed one. In the fall of 2015, three separate groups published papers nearly back-to-back reporting on experimentally closing both loopholes. I got NIST and Austria spot-on but missed a group in the Netherlands.[28] (See Figure 3.8.) And so there we have it. Einstein's three horrors of the quantum theory have been ruled *in* – to immense precision over vast distances. Most sensible physicists do not debate these results. The three weird features of quantum theory are neither philosophy nor religion – they are reality – we have to accept them. Not only do we have to embrace them, but also now we should put them to work. That is quantum technology.[29]

28 For a history of Bell-test experiments, see, https://en.wikipedia.org/wiki/Bell_test_experiments#Hensen_et_al.,_Giustina_et_al.,_Shalm_et_al._(2015):_%22loophole-free%22_Bell_tests.

29 Here is an excellent video taken of me, at the National Academy of Sciences, discussing this topic, *Quantum Technology: Prof. Jonathan Dowling*. www.youtube.com/watch?v=IHNV_CzCeA0&feature=youtu.be.

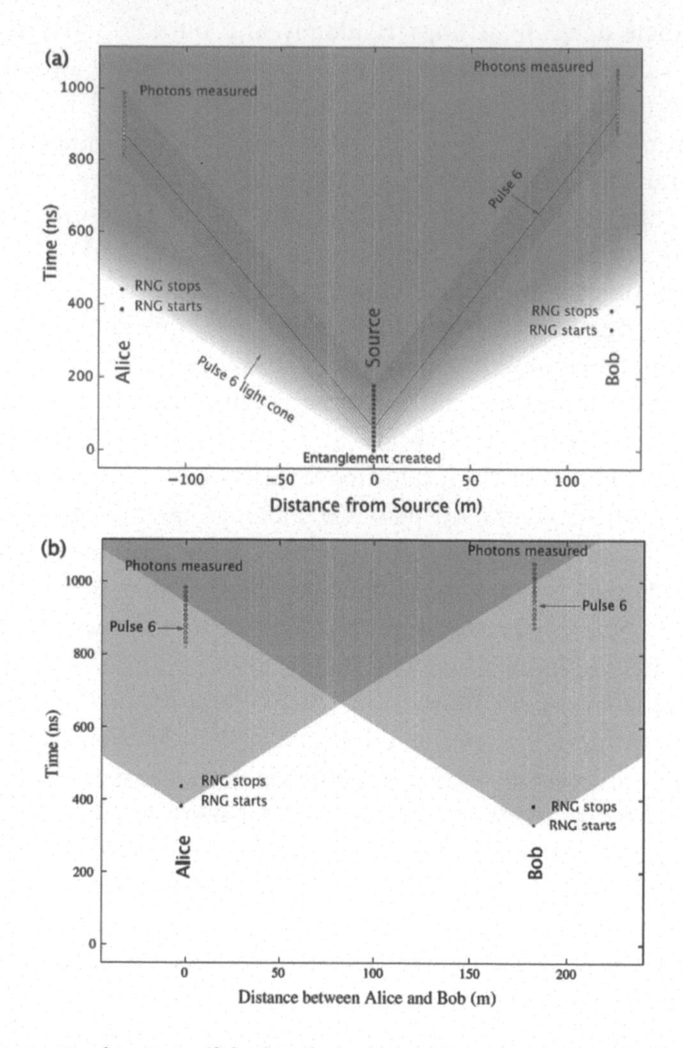

FIGURE 3.8 A schematic of the loophole-free Bell test done at NIST in Boulder, Colorado. Firstly, they used detectors with greater than 83% efficiency to close the detector loophole. Then they separated the source and detectors over substantial distances of hundreds of meters. The blue regions indicate signals that are moving at the speed of light. Firstly, Alice and Bob's choices of the measurement basis must be carried out before any gremlin at the source can reach gremlins at their remote locations by mobile phone. This speed limit is indicated by noting Alice and Bob are below the two blue lines emanating from the source (a). The RNGs are random number generators. The gremlin signals can only travel along those outermost blue lines. The figure indicates that the

FIGURE 3.8 (Cont.) gremlin signals, moving at the speed of light, arrive at Alice and Bob's locations after they choose their detector settings. Now consider the two blue lines emanating from Alice's position — any mobile phone signal sent by a gremlin on Alice's station must travel along these two blue lines (b). Now notice the two blue lines coming from Bob's station. Any mobile phone signal from a gremlin at Bob's station must move within these blue regions. The fact that Alice makes her measurement and computes her result below Bob's blue line means that no two gremlins on either Alice's or Bob's positions can communicate with each other before this measurement occurs. This shows that the sprites on Alice and Bob's detectors cannot conspire to skew the data unless they can communicate faster than the speed of light.[30]

TWO'S COMPANY – BUT THREE'S A CLOUD

In addition to the four, two-photon, Bell states, discussed above, there are two classes of entangled three-party states where Alice, Bob, and Charlie share three entangled photons among them. These states serve as primitives for additional network capability such as quantum voting, anonymous quantum broadcasting, distributed quantum sensors, and quantum cloud computing. We will discuss these protocols in what is to come, but in this chapter, I want to lay out the groundwork for the properties of the most-used three-photon-entangled states so that we can refer back to them later. Three-party states naturally form a two-dimensional network, like the internet, that two-party states miss out on. The two states we shall discuss now are the Greenberger–Horne–Zeilinger (GHZ) state and the *W*-state.

Three-photon entangled states, which can be distributed to three distant parties, Alice, Bob, and Charlie, can be used to make-up the nodes of a two-dimensional network where the three parties are located on the vertices of a triangle. There are two fundamentally different three-photon states. The first we'll discuss is the GHZ state, named for the American physicists Daniel Greenberger and Michael Horne, and the Austrian physicist, Anton Zeilinger.[31] The state was initially conceived as

30 The figure is the property of the U.S. Government and not subject to copyright; Strong Loophole-Free Test of Local Realism, Lynden K. Shalm et al., Phys. Rev. Lett. 115, 250402 – Published 16 December 2015; https://journals.aps.org/prl/abstract/10.1103/PhysRevLett.115.250402.

31 For more on Danny Greenberger, see https://en.wikipedia.org/wiki/Greenberger%E2%80%93Horne%E2%80%93Zeilinger_state.

a new state that could put even further constraints on the Einstein local-hidden-variable theory. It is a three-photon state shared by three parties, Alice, Bob, and Charlie. It has the form, $|\text{GHZ}\rangle_{ABC} = |H\rangle_A|H\rangle_B|H\rangle_C + |V\rangle_A|V\rangle_B|V\rangle_C$. The interpretation is that Alice, Bob, and Charlie *all share either* three H-polarized photons, or *they all share* three V-polarized photons, but cannot know which is which. (Perhaps it is better to say, as before, the three share neither all horizontal photons nor all vertical.) These states are sometimes called Schrödinger cat states or N00N states. The first odd thing is this. Let us say Alice measures her photon using her polarization analyzer. If the result of her measurement is H, then Bob's and Charlie's photons also instantly collapse to H, so when they all three measure their polarizations, they all three get H, regardless of the distance of their separation. If instead, Alice's measurement produces a V, then the other two photons collapse to V. Thus, Bob and Charlie will also get a V when they measure. It does not matter who measures first. Upon measurement in the H–V basis, all three of them always get an H, or all three get a V. There is a 50–50 probability for either outcome. If Alice makes a measurement using a nondemolition analyzer, then the GHZ state collapses into either $|H\rangle_A|H\rangle_B|H\rangle_C$ or $|V\rangle_A|V\rangle_B|V\rangle_C$, and so the entanglement is entirely lost.

Now, like before in the Bell test, we have a new type of analysis. Alice, Bob, and Charlie each make a single measurement, not in the H–V basis but the ±45° basis. They record their three outcomes and then plug them into a new and improved Bell tester. If the Bell tester spits out the number one, Einstein's classical hidden-variable theory is right, and if it spits out a negative one, then the quantum theory is correct. The improvement is that you can distinguish a positive one from a negative one with a single measurement, whereas before – to separate less than two from greater than two – required many measurements. As we now expect, in all experiments, the Bell tester spits out a negative one, once more reaffirming that the quantum theory rules over Einstein's.

While we cooked up this state for testing quantum theory, it now has applications to a host of quantum-internet protocols. Some of these include making accurate atomic clocks, synchronizing atomic clocks in a network, and something called quantum-anonymous broadcasting. In broadcasting, any one of the three parties, Alice, Bob, or Charlie, may make a bid on a pricey piece of art in an auction, and everybody at the auction will know what the bid was. But nobody can tell – aside from the bidder – who made the bid.

Now we come to the *W* state. I don't know why it is called that but let's say because it's weird. It is also a three-photon-entangled state. We can show that there are only two types of unique three-photon states, the GHZ state, and the *W* state. The latter is defined by $|W\rangle_{ABC} = |H\rangle_A|V\rangle_B|V\rangle_C + |V\rangle_A|H\rangle_B|V\rangle_C + |V\rangle_A|V\rangle_B|H\rangle_C$. There are only three photons, but they are superimposed in three ways. Either Alice has an *H*, and Bob and Charlie both have V's, or Bob has an *H*, and Alice and Charlie both have *V*'s, or Charlie has the *H*, and Alice and Bob both have *V*'s. Nobody can say for sure who has what. It follows that if all of them set their analyzers to measure in the *H*–*V* directions, if Alice gets the *H*, then she knows that Bob and Charlie both will measure *V*s. But if Alice gets a *V*, then she only knows that either Bob got a *V* and Charlie got the *H*, or Bob got the *H* and Charlie got a *V*, but she cannot tell which is which.

We can use this property to do the following neat trick. Suppose Alice's photon is not measured but rather entirely lost, such as might happen in a lossy fiber. We can show that if this happens, then Bob and Charlie's photons collapse into the entangled Bell state $|\Psi^+\rangle_{BC} = |H\rangle_B|V\rangle_C + |V\rangle_B|H\rangle_C$, with a probability of 2/3, or it collapses into the unentangled state $|H\rangle_B|H\rangle_C$ with an expectation of 1/3. So unlike the GHZ state, where measuring a photon always loses all the entanglement, with the *W*-state there is a 66% chance you'll end up with the two remaining photons still entangled.[32] Now consider a larger *W*-state shared between four photons that we send to Alice, Bob, Charlie, and Doug,

$$|W\rangle_{ABCD} = |H\rangle_A|V\rangle_B|V\rangle_C|V\rangle_D + |V\rangle_A|H\rangle_B|V\rangle_C|V\rangle_D + |V\rangle_A|V\rangle_B|H\rangle_C|V\rangle_D + |V\rangle_A|V\rangle_B|V\rangle_C|H\rangle_D.$$

This time, if Alice's photon is lost, the thing collapses back to the three-photon *W* state, $|W\rangle_{BCD} = |H\rangle_B|V\rangle_C|V\rangle_D + |V\rangle_B|H\rangle_C|V\rangle_D + |V\rangle_B|V\rangle_C|H\rangle_D$, with a probability of ¾ or into the unentangled state

32 A brief review of the *W*-state is here, https://en.wikipedia.org/wiki/W_state.

$|H\rangle_B|H\rangle_C|H\rangle_D$. Now, if Bob also loses a photon, Charlie and Doug still have an entangled pair of photons with a total probability of $3/4 \times 2/3 = 1/2$. So a quantum network with a large shared W-state has the robustness to photon loss if the goal of the network is for each node to share entanglement with all the other nodes.

SPEAKING THE UNSPEAKABLE

I have quickly covered the evolution of quantum foundations from about 1935 up until 2018. Remember – in 1984 it was considered crackpot stuff – I'd never get a job! Now we know these weird predictions of quantum theory are right, just as well as we can understand that *anything* is true. We have verified them to immense precision over vast distances. A measurement on a photon here changes the outcome of a measurement on a photon 1,200 km away. Quantum unreality is a reality. So are uncertainty and nonlocality. The things Einstein so loathed about quantum theory are measurable features of our natural world. When I lecture on this stuff, often I'm asked, "Do you *really* believe this?" My response now is, "If you want belief, go see a rabbi or a priest." I'm also asked, "But what does it mean?" My response is, "If you want meaning, go see a philosopher." I'm teaching this material in a graduate course called *The Quantum Internet.* I have an engineering student in the class, but most of the rest are physics students. When I get to this part of the lecture, the physicists are flummoxed. I ask the engineer, "William? What do you think about all this?" His response was firm and instantaneous (and very engineering like), "I don't care *what* it *means* – I only care that it *works!*"

A scholar of the foundations of quantum theory is like the Rodney Dangerfield of theoretical physics. An American comedian, Dangerfield was known for his New York accent, pop-eyes, jokes about his wife, and his catchphrase, "I don't ever get no respect!"[33] And yet we have this other field of physics called superstring theory that was supposed to be the only game in town. For thirty years, superstring theorists were the superstars of the physics theory world. They wrote popular books on the topic, cornered

33 "A full ten minutes of Rodney Dangerfield doing stand-up and having some laughs with Johnny Carson. Originally aired 1 August 1979 on *The Tonight Show.*" www.youtube.com/watch?v=jrFgD9-l390.

the market for all the government funding, and appeared on talk shows and television shows promoting an approach that has not a single shred of laboratory evidence to support it whatsoever. Over the same 30 years, we have measured quantum foundation's strange predictions to incredible precision and over vast distances. There are infinitely more data to support the predictions of quantum foundations than there are to support the superstring theory. And yet, "we don't ever get no respect."

One of the premier places in the world to do theoretical physics is the Perimeter Institute in Canada. A few years back, in a night-of-the-long-knives moment, all but one of the quantum foundations scientists were summarily sacked, so the upper administration could replace them with even more superstring theorists. Superstring theory has a big problem. It makes predictions of new particles that should have been seen at the giant atom smasher at CERN and in a host of other very expensive experiments. There is no sign anywhere in any experiment of any of these predictions. The simplest and most beautiful version of superstring theory made predictions that are wrong. Now the superstring theorists, desperately clutching at ropes, are tweaking the theory that was too beautiful to be wrong, making it uglier and uglier so that it does not contradict the data pouring in. Talk about moving goalposts! It is as if they promised Prince Charming that he would meet Cinderella at the ball – but then they produced her stepsister instead! For more on this trouble with physics, I recommend four popular books: *The Trouble with Physics*, by Lee Smolin; *Not Even Wrong*, by Peter Woit; *Farewell to Reality*, by Jim Baggott; and *Lost in Math*, by Sabine Hossenfelder.[34] The last book is a real hoot.

Meanwhile, thanks in part to a 2003 paper by Australian physicist Gerard Milburn and me, "Quantum Technology – The Second Quantum Revolution," the foundations of quantum theory has morphed into quantum technology, as we predicted it would.[35] (I should have trademarked

---

34 Smolin, Lee. *The Trouble with Physics*. Penguin Books, 2008. www.worldcat.org/oclc/942592494; Woit, Peter. *Not Even Wrong*. Vintage Press, 2011. www.worldcat.org/oclc/1004571212, Baggott, Jim. *Farewell to Reality*. Pegasus Books, 2014. www.worldcat.org/oclc/1041414700; Hossenfelder, Sabine. *Lost in Math*. Basic Books, 2018.

35 Dowling, Jonathan P., and Gerard J. Milburn. "Quantum Technology: The Second Quantum Revolution." *The Philosophical Transactions of the Royal Society of London* 361 (2003): 1655–1674. doi:10.1098/rsta.2003.1227. http://rsta.royalsocietypublishing.org/content/roypta/361/1809/1655.full.pdf.

that title. It is used all the time now, but not everybody cites us, or sometimes they cite Milburn and not me. "I don't ever get no respect." If you Google the phrase "the second quantum revolution" before 2003, you get zero hits. If you Google it now, you get 30,000 hits.) If we go back to chapter one, we are in a similar position now as the field of electromagnetism was in the mid-1800s. Faraday and Maxwell had all these ideas about invisible force fields filling all of space-time – who could believe *that?* It was not until the experiments of Hertz and Marconi – 50 years later – that people began to believe. When I teach these theories to my undergraduate engineering students, they are also flummoxed. It took 50 years from when Maxwell first wrote down his four famous equations until the engineers figured out how to make generators and electric motors and power a grid from them. I suppose many of them still don't care what it means – they only care that it works. Einstein first pointed out these weird properties of quantum theory in 1935. It took us 83 years to determine that they are real, they are here to stay, and they are useful for lots of things, such as the quantum internet. However, we are just now developing university programs in quantum engineering. What we need are engineers trained in enough quantum theory to understand how it works, if not why it works. It is a goal of this book to be a primer for quantum optical engineering for the quantum internet. Let us physicists stop arguing over what these predictions of quantum theory mean and allow the engineers to put them to work. In the words U.S. Army General George Patton, it's time for physicists to either "… lead, follow, or get the hell out of the way!"

CHAPTER 4

Quantum Networks: The Building Blocks

[1]

In the case of all things that have several parts and in which the whole is not, as it were, a mere heap, but the totality is something besides the parts, there is a cause of unity; for as regards material things contact is the cause in some cases, and in others viscidity or some other such quality.

Aristotle[2]

1 www.shutterstock.com/image-photo/kids-play-room-child-hard-hat-105983540.

2 Edited by Jonathan Barnes, *Complete Works of Aristotle, Volume 2: The Revised Oxford Translation*, 1650. Princeton, NJ: Princeton University Press, 1984. www.worldcat.org/oclc/8344807.

SINGLE-PHOTON GUNS

In the previous chapters, we outlined the importance of having a reliable source of single photons on-demand. While single photons are not entangled, as it takes two to tangle, they do display two out of three of Einstein's no-noes – unreality and uncertainty. We can use these two features as a resource in quantum cryptography, some types of optical quantum computing, and quantum sensing. I'm sitting here in my office at the University of Science and Technology of China, drinking a double espresso, after just having had lunch with my colleague Chaoyang Lu. In the lunch discussion, he asked me, "Why are people not putting more effort into developing reliable single-photon sources? They are very important." Indeed, they are and some of the best single-photon sources in the lab here in Shanghai. What do we want a single-photon gun to do? Well, it should be like a gun. You pull the trigger, and you get one and only one photon out every time and, you never get zero photons out, and nor do you ever get more than one photon out. A gun that mostly shot blanks would be good for playing Russian Roulette, but not for much else. We also need that each photon fired is indistinguishable from any other photon by any means, except by the time of firing. By that, I mean that each photon should have the same polarization, wavelength, pulse shape, etc., like every other photon. This property will be critical when we look at two-photon interferometers. And finally, when we shoot the gun, the bullet should come straight out of the end of the barrel and not misfire and shoot the shooter in the foot. What do we have that does this? Nothing precisely, but there are three different types of guns in common use.[3]

QUANTUM-DOT SINGLE-PHOTON GUNS

Here at the University of Science and Technology of China, they have one of the best single-photon guns. In my after-lunch discussion with Lu today, I learned that their best gun fires a blank 90% of the time, single photons 9% of the time, and the probability of two photons per shot is 1% of the time. That is state of the art. Just a year or so ago, the record was blanks 99% of the time and single photons 1% of the time. If

3 For more on single-photon sources, see https://en.wikipedia.org/wiki/Single-photon_source.

you make a plot of this particular solid-state on-chip single-photon gun performance as a function of time, the improvement toward a perfect single-photon gun is growing very rapidly.

How are these guns made? (Very carefully.) The designs differ a bit from place to place, but the idea is mostly the same. To make sure you have a single photon, you make sure you have a single atom. A single atom, when excited to an upper state, will eventually decay back down to its ground state and emit one and only one photon, because there is one and only one atom. However, in this process, called spontaneous emission (the atom spontaneously emits a photon in a certain time window), if the atom is sitting in empty space, the photon can come out in any random direction. Not a very good gun if it shoots the shooter in their foot or even their head! Also, it is hard to hold an atom in empty space without it falling to the ground.

The way around this problem is to put the atom into an optical cavity, which is a device that traps the photon between two mirrors. The enclosure serves two purposes; one is to hold the atom still, and the other is to direct the photon out in a single direction like the barrel of a gun does. The final step, to tailor this gun for our needs, is to replace the real atom with an artificial atom called a quantum dot.[4] A quantum dot is an engineered atom made out of semiconductor material that you design to have the properties you want it to have. In a real atom, you are stuck with the properties that Mother Nature gave you. These devices are called quantum-dot single-photon sources, and they are some of the best around.[5] I show a schematic of such a device in Figure 4.1. The quantum dot is grown into a multilayer structure of semiconducting material. Then most of the material is etched away leaving a single micropillar containing the quantum dot between the two mirrors. (The micropillar resembles a stack of pancakes.) It can be shown that when the dot emits a photon, it is very likely to go straight up – the barrel of the gun – and not shoot you in your foot. The stack of pancakes acts to guide the photon to the top of the stack and then out.

---

4 This link takes you to a page on quantum dots, https://en.wikipedia.org/wiki/Quantum_dot.

5 Here's how to make a single-photon gun out of a quantum dot, https://en.wikipedia.org/wiki/Quantum_dot_single-photon_source.

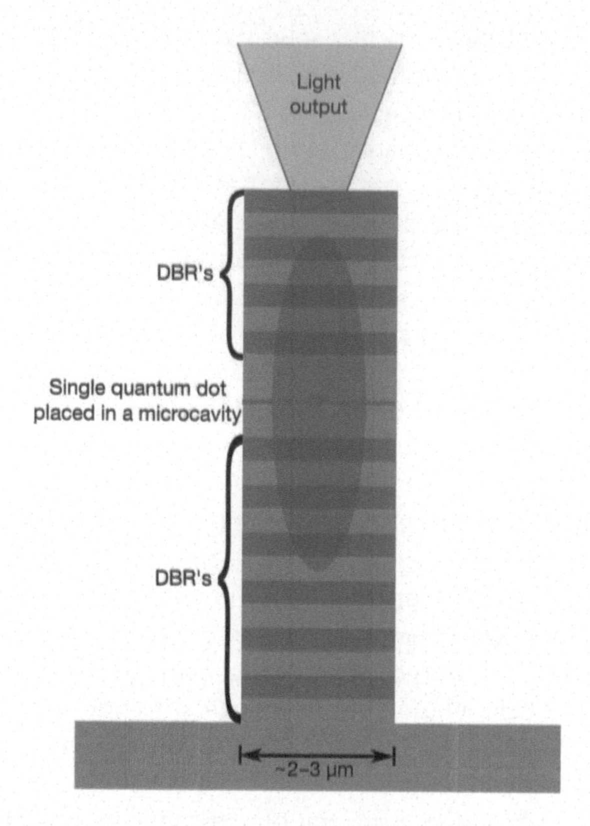

FIGURE 4.1 Schematic structure of an optical microcavity with a single quantum dot placed between two layers of distributed Bragg reflectors (DBRs), which are types of mirrors. This structure works as a single-photon source. We design the microcavity and the mirrors to direct the photon out of the top.[6]

This thing is cylindrically shaped and very small; as you can see in the figure, it is a micron in diameter, which is about a hundred times smaller than the diameter of a human hair. The quantum dot can be excited either electrically or optically with a laser pulse. We prefer the latter method, since we clock the laser pulses. In the ideal case, a single photon would come out each nanosecond. A nanosecond is one-millionth of

6 This graphic, https://en.wikipedia.org/wiki/Quantum_dot_single-photon_source#/media/File:Optical_Microresonator.png, is licensed under the Creative Commons Attribution-Share Alike 4.0 International license, https://creativecommons.org/licenses/by-sa/4.0/deed.en, and was created by Wikipedia user Phezack.

a second, that is, a million single photons per second – a respectable rate. However, exciting the quantum dot with a laser at the same wavelength as the photon that the dot will emit comes at a cost. Sometimes a stray photon from the exciting laser itself gets caught in the gun, and the gun can emit two photons at once – one from the atom and the other the stray photon from the laser. After decades of work, this probability of getting two photons per shot is now much less than 1%.

Additionally, while the optical cavity micropillar is a big improvement from an atom floating around in the air, it is not perfect. You are now unlikely to shoot yourself in your foot or your head, but like with a sawed-off shotgun, the single photons come out still with quite a spread, and many don't hit the target. In this case, the target is an optical communications fiber that carries the photons away to do their duty somewhere else. This inefficiency is why the probability of getting no photon per shot is 90% – about 90% of the time the photons miss the target and go flying off at some angle. The remaining 10% hit the target, go into the fiber, and can be put to work. After some more years, I expect this ratio will reverse, and 10% will be lost and 90% will hit the target. The improvement in this technology is very rapid.

In Figure 4.2 we show a plot of the properties of such guns as a function of photon indistinguishability (vertical axis) and the probability of the gun firing at least one photon (horizontal axis). The best single-photon gun would be up in the top right corner.

One of the key advantages of the quantum-dot single-photon guns is that we can manufacture them cheaply and in mass quantities – they fit onto a compact circuit board. In the quantum internet, you will need millions of these things distributed all over the Earth, and hence they need to be small and cheap and run on low power.

SORT-OF-SINGLE-PHOTON GUNS AND MAGIC CHINESE CRYSTALS

Rather than listing single-photon guns here in a chronological order, I'm giving them in the order of their performance from best to worst. But even the worst sources have a commercial application. The magic Chinese crystal photon guns have a long history. Since we can use these guns to generate entangled photon pairs, I will spend a bit of time discussing them here. There is a process in optics called *downconversion* where if a photon hits a special type of crystal, it downconverts into two

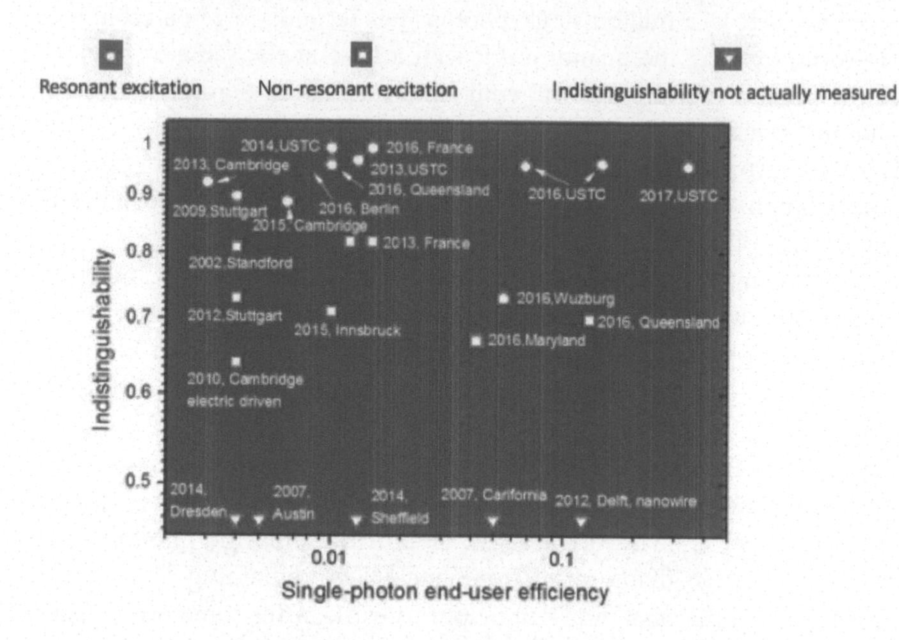

FIGURE 4.2 The plot of quantum dot single-photon gun performance. Each data point is a reported experiment labeled by group and year. The vertical axis is the photon indistinguishability. When this quantity is one, all photons are exactly the same. The horizontal axis is the probability that when you pull the trigger at least one photon is fired, where the far right is the best at 100%. So, the best guns would be in the top right corner.[7]

photons, each double the wavelength of the pump photons. I spent a lot of time telling you in chapter two that there is no such thing as half a photon, but that is not what we are doing here. The magic crystal absorbs the incoming photon, and then *two new photons* are emitted, each with double the wavelength of the original. (Actually, it does not have to be double, it just has to be the wavelength of one daughter photon, and the wavelength of the other daughter photon must add to be half that of the mother photon.) I'm not splitting the mother photon – it gives birth to two twin daughter photons and then vanishes. If the magic crystal is sufficiently magic, then the two twin daughter photons are identical. I illustrate the process in Figure 4.3.

7 Figure provided by Chaoyang Lu at the University of Science and Technology of China with his full permission to reuse and edit it for any purpose.

The laser pump contains the mother photons (in blue). The daughter photons are labeled signal (s) and idler (i) for historical reasons. The green box is the magic crystal.[8] Somewhat technically, the momentum of the daughter photons adds up to the momentum of the mother photon, as does the energy, and so momentum and energy are conserved.

The good news is that *when* the mother photon gives birth, *she always has twins*. The bad news is that the mother photons *are much more likely* to just pass through the crystal without interacting with it, and no daughter photons are emitted.[9] So if there is a daughter photon in the signal beam, you are *sure* there is a twin in the idler beam, but most of the time, you get out nothing at all. (There is a much smaller probability that you get identical quadruplets with two daughters in the signal arm and two daughters in the idler arm.) How do you make a single-photon gun out of this device? The idea is to put a photon detector into one of the

---

8 The crystals are not magic. However, they seem like magic. The proper name for them is a type of nonlinear optical material. As told to me by Yanhua Shih at the University of Maryland, here is their story. Back in the period of 1966–1976, China underwent the cultural revolution. Universities were shut, and university professors were sent to work on farms. The Red Guards roamed the universities enforcing these policies until they came to a bunch of chemists in a shed. The making of these magic optical crystals is something of a black art. You start with a vat of hot steaming goo and then seed the goo with a bit of crystal on a string. Then slowly you cool the goo and stir it and if you do all this just right the tiny sand-grain-size crystal grows into a perfect crystal about the size of a small box of matches. If you don't do it right, it grows into a disorganized noncrystalline mess, and you start all over again. The Red Guards arrived in this shed behind the chemistry department and saw these sweaty professors barefoot wearing shorts and head bandanas but no shirts stirring this immense and boiling hot vat of goo. The Red Guards stared at them for a while and then said, "That looks like hard work. You stay here and keep doing that." While the rest of the faculty worked the farms for ten years, these guys spent ten years perfecting the growth of the magic Chinese crystals.

9 Irish joke as told to me in 1982 by the Nobel Laureate Paul Adrien Maurice Dirac at a reception in his honor at the University of Colorado. I went up to Dirac and apologized for interrupting his lecture (see my previous book). He seemed to have forgotten all about it. He asked me where I was from, and I explained that I was a dual Irish and American citizen. Then, with no warning, Dirac told the following somewhat inappropriate joke.

"As is typical in Irish Catholic parishes, each Sunday evening the parish priest would visit a family in the parish for a free dinner and to check up on their mortal souls. When the new family moved into the parish, the priest made sure to schedule such a dinner. When he arrived, he was introduced to the mother and the father and their ten children. Now, having ten children is not unusual for an Irish-Catholic family, but the amazing thing was that the children were *five sets* of *identical* twins. The priest exclaimed to the mother, 'Do you mean to tell me that you *always* had twins!?' To this, the mother replied, 'Oh *no* father – *most* of the time we had *nothing* at *all*!'".

Dirac then gave me the faintest hint of a smile and walked away. That was the last time I ever saw him as he died two years later. It was not until 20 years later that I understand that his off-color joke was a perfect analogy of the process of spontaneous parametric down conversion. Every once in a while, you get twins – but *most* of the time you get *nothing at all!*.

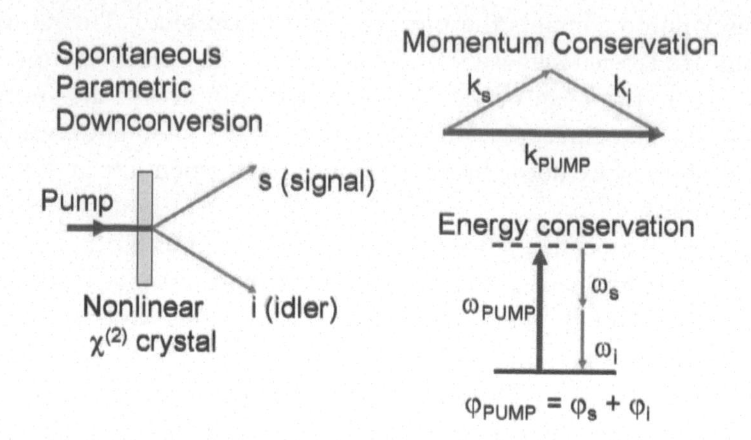

FIGURE 4.3 The process of downconversion where a mother photon (blue) is absorbed by the magic crystal (green), which then emits two identical daughter photons (red).

arms – let us put it in the idler arm. Every once in a while, the photon detector will go click. When it does, it means that you have detected one of the twins, and you are sure that the other is in the signal arm. We call this process heralded single-photon generation – the detection of a photon in the idler heralds the presence of a photon in the signal. We sacrifice the idler photon, but you can be sure there is a single photon in the signal. As you run this thing each time when the idler detector clicks, a signal photon emerges. Due to the special properties of the crystal, all photons in the signal are identical to each other. So how does this machine stack up to a perfect single-photon gun?

A perfect gun should emit one photon every time we pull the trigger, each photon must be identical, and the gun should shoot straight. Well, for the first criterion, there is no trigger. The thing is like a bad automatic machine gun – firing blanks most of the time – and a photon pair only once in a while. Instead of a trigger, we have a herald. When the gun fires a pair, one photon, the idler, is used to herald a photon in the signal. Also, while rare, sometimes the gun will shoot out two photons in the signal instead of one. But most of the time it shoots out nothing in the signal. The critical thing that makes it work at all is that, due to the heralding, when it does shoot something, you know it. The gun also has quite a spread of the photons, more like a sawed-off shotgun. You are unlikely to shoot yourself in the foot, but the photons come out in a broad cone, as shown in Figure 4.4. With the

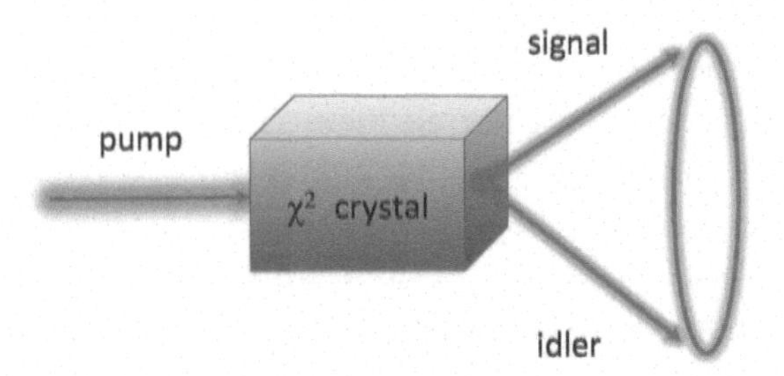

FIGURE 4.4 A 3D schematic of the downconversion process. The twin daughter photons come out on opposite sides of a circular cone. You have to place a mask in front of the cone with two diametrically position pinholes to make sure the photons in the signal come out in a tight beam.

cost of losing precious photons, what you typically do is select two opposite points on the cone and put an opaque mask in front of the cone with small pinholes diametrically positioned on the mask. Then you get something that looks more like Figure 4.3 where the idler goes one particular way, and the signal goes the other, each in two narrow beams. When the idler detector clicks, then you have a single photon in the signal beam that you can feed into an optical fiber for further use.

While this does not seem like a very good gun compared to the quantum dots, it has many advantages. You can buy these magic crystals now from many sources, and students in a lab with lenses and mirrors can do the experiments without having to figure out how to put everything on a chip. These magic crystal devices have been the work-horses of single-photon guns since about 1990. Only now are the microfabricated quantum dots catching up in performance. However, as we shall see, these magic crystals have more tricks up their sleeves. The twin daughter photons are entangled!

PSEUDO-SINGLE-PHOTON GUNS

For applications such as quantum cryptography, for 30 years we have used attenuated laser pulses instead of true single-photon sources. What on earth is an attenuated laser pulse? Recall the requirement of photons guns. They must produce one and only one photon per shot and never fire blanks. The photons must be indistinguishable, and they must always come out the barrel of the gun in a tight beam and not go every which way. As we have seen, none of the photon guns discussed so far meet all those criteria. In this section, we will discuss yet another photon gun that does not meet all these criteria either. It is called an attenuated laser pulse. The light that comes out of a laser is as classical as you can get. It almost obeys the Maxwell wave equation for classical light waves, so long as the laser is sufficiently intense. However, there is no such thing as classical mechanics, so in some regimes, a laser manifests quantum behavior. The light coming out of a laser pointer contains an average of 1,000,000 photons per second. At this intensity, the photon nature of the light is hardly discernable. To make a pseudo-single-photon gun, you must attenuate the laser, pass it through an absorbing filter, so that the output contains, on average, a single photon per second (or less). This type of photon gun has the same problem as the other ones. Much of the time, when you pull the trigger, you get nothing. If the average number of photons emitted is one per second, then on average, you will get one photon per shot. However, that is just the average, and sometimes you will get two photons per shot, but most of the time, you will get zero per shot.

As we will see, when we discuss quantum cryptography devices in the next chapter, those events when two photons come out are perilous, as an eavesdropper can snag the extra one and learn something about the secret cryptographic key you are trying to set up. We have ways of thwarting that, and laser pulse technology is 30 years ahead of the other two types of single-photon guns we discussed above. For this reason, it is widely used, particularly for quantum cryptography. The entire Chinese quantum cryptography system, which now spans much of their country, relies on these laser pulses. The main advantage is that we can make them fire very fast, and that increases the rate the users can accumulate secret key. Besides, there is a way to mitigate the problem of extra photons, whereby Alice randomly sends a bright pulse to Bob instead of a weak key-bearing pulse. These bright pulses are called decoy states.[10] Deploying decoy states is analogous to a situation where Alice and Bob are in a dark cavern, communicating with very weak laser pulses, when suddenly and at random Alice turns on a bright searchlight. The eavesdropper is then sure to be caught. QKD using attenuated laser pulses

---

10 For more on decoy states, see https://en.wikipedia.org/wiki/Decoy_state.

with decoy states is provably unbreakable, and this is the scheme the Chinese are using both on their landlines and their space links.

In Figure 4.5, I show a comparison of the output of three-photon guns: an idealized quantum dot gun, a magic crystal gun, and the attenuated laser pulse gun. The horizontal axis labels the photon number that comes out of the gun: zero, one, or two. The vertical axis labels the probability the gun shoots that number. In all cases, I have taken the average number of photons emitted by the source to be one. A perfect photon gun would have one dot located at one on the horizontal axis and at one on the vertical axis. The idealized quantum dot (blue dots) performs the best, with a high probability of nearly one (100%) of emitting a single photon and a low probability of 0.01 (1%) of emitting either nothing or two photons. That is why there is a race to perfect these quantum-dot guns. The magic crystal gun (orange dots) and the laser pulse gun (green dots) perform comparably badly. They both have a much higher probability than the quantum dot gun of shooting a blank (40% or more) or shooting two

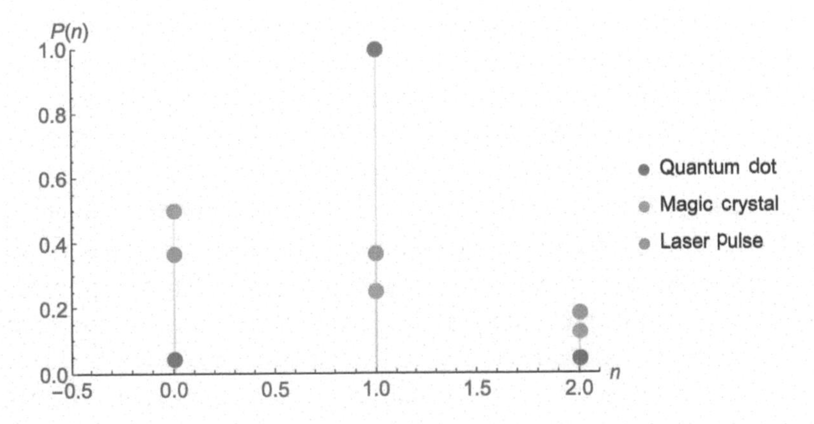

FIGURE 4.5 A comparison of the performance of the idealized quantum-dot guns (blue dots), the magic crystal gun (orange dots), and the laser-pulse gun (green dots). The vertical axis is the probability of the gun firing zero, one, or two photons. A perfect single-photon gun would only have a single dot at one on the horizontal axis and one on the vertical axes. The idealized quantum dot comes closest. The other two have a higher probability of shooting blanks or two photons at once.

bullets at once (around 20%). As mentioned, for quantum cryptography, a gun that shoots two bullets at a time is susceptible to attack from an eavesdropper who can steal one of the bullets and learn something about the cryptographic key.

ENTANGLED-PHOTON GUNS

One of the primary components of the future quantum internet will be a small chip-sized device that spews out entangled photons to users on the network. The entangled photons will then be transferred into quantum memories so that the two users share quantum entanglement between them stored in the memories. Once two users, say, Alice and Bob, share a large number of entangled particles, there is a host of protocols that they can use these entangled pairs for. Some are quantum cryptography, distributed quantum computing, teleportation, long-baseline telescope arrays, clock synchronization, and networks of quantum sensors. There are a several communication protocols that required some sort of shared entanglement. Examples are secret sharing and anonymous broadcasting. Someday there will be a Department of Entanglement whose job it is to keep the entanglement coming just like the Department of Energy has the job to keep the lights on. We can imagine millions or billions of these little entanglement-spewing machines all over the network – constantly providing the users with shared entanglement. As discussed in the previous chapter, there are three types of entanglement of interest, the entangled photon pairs, called Bell pairs, and the two types of entangled photon triplets, the GHZ state and the *W*-state. Each of these has its uses. We'll start with the two main approaches to produce Bell pairs. These approaches mimic two of the technologies discussed in the section on single-photon guns, but now they are double-photon guns where the two photons are entangled. The two main approaches are again quantum dots and magic crystals. The magic crystal approach is far more developed than the dot approach, but the dots are coming along. To be consistent, I will start with the dots.

QUANTUM-DOT ENTANGLED-PHOTON GUNS

The simplest example of such a thing is a dot that fires out entangled pairs of photons. The concept goes back to the 1970s when the first Bell tests were done with cesium atoms. These atoms were held in a gas chamber and illuminated with a weak light source, which excited the ground-state electron to an upper state. Eventually, that electron would come back down by one of two nearly identical paths, as shown in Figure 4.6.

Due to quantum unreality, the electron takes both paths down simultaneously and emits the entangled photon state, $|H\rangle_{\text{Alice}}|V\rangle_{\text{Bob}} + |V\rangle_{\text{Alice}}|H\rangle_{\text{Bob}}$, as shown. The problem with this source, as is the problem with atom-based single-photon guns, is that the atom fires the entangled pairs always in opposite directions but rarely the two directions where Alice and Bob are sitting. The idea of the entangled-photon source in a dot is very similar to the single-photon source in a dot. We can engineer the dot to have the level structure of Figure 4.6. We choose what wavelengths you would like and not what Mother Nature has given you. We put the source in the same micropillar structure of Figure 4.1, which guarantees that most of the time that the entangled photon pair will emerge from the top of the micropillar. The dot is excited using a very fast pulsed-laser beam, and the thing can emit

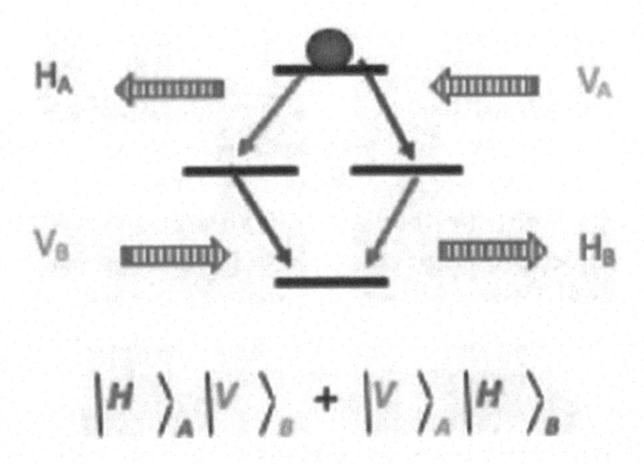

FIGURE 4.6 Entangled photons emitted by an excited cesium atom. The electron is excited to the uppermost level (round dot). After a while, it decays to the lowermost level. If it takes the right-hand route down, it emits a V-polarized photon toward Alice and an H-polarized photon towards Bob. If it takes the leftmost route down, it does the reverse. Due to quantum unreality, it takes both routes down simultaneously and emits the entangle photon state, as shown.[11]

11 I drew this figure myself in PowerPoint, and it shows.

millions of entangled photon pairs per second, whereas, in the old cesium atom experiments, you were lucky to get a few tens of photon pairs per second.

You continue along this line and make dots in micropillars that emit three entangled photons on-demand, either in the GHZ-state or the W-state. In the quantum internet, most of the nodes will contain guns like this that spew out entangled pairs or triplets for users to exploit at the receiving nodes. The dots are not perfect yet, but they improve every year, and they have the quality of being small and compact things on chips. Currently, the problem with this source is that it must be cooled to near absolute zero to work well. The cost, size, and power consumption of the refrigerator is the main drawback to commercialization. Someday we'll learn to make these dots so they operate at room temperature, but in the meantime, the fridges are getting more compact, cheaper, and easy to use.

However, two more issues are hampering the development of entangled-photon dot guns. The first is that it is very difficult to make the two pathways down in the dot in Figure 4.6 identical. For cesium atoms, this is easy, because cesium atoms floating in empty space have a nice clean level structure. For the dots, they are immersed in the semiconductor material that makes it difficult to make the paths the same. If they are not exactly the same, instead of an entangled photon pair, you get an unentangled pair. That is back to classical coin-flipping. Half the time, a *V* goes to Alice and an *H* to Bob, and the other half the reverse. The other problem is that, unlike cesium atoms, no two dots are exactly alike. So if you want to make an array of these dots all spitting out identical entangled photon pairs, you have to tune each dot separately by adjusting magnetic or electric fields applied to the dot so that they are all identical. It is difficult to do this with many dots on a chip. That is why, for years, the crystals have been the workhorses for entangled-photon sources.

ENTANGLED SORT-OF-PHOTON GUNS AND MAGIC CHINESE CRYSTALS

The same crystals that we used to make heralded single-photon guns, as discussed above, produce entangled photon pairs naturally. They are much more like the cesium atom, so you do not run into issues such as with the dots. The trick is to take the downconverters (shown in Figures 4.1, 4.3, and 4.4), and cut them at a slightly different angle than is used for the single-photon guns. After we do this, the colorful cone of entangled photons splits into two cones, as shown in Figure 4.7.

The crystal is made in such a way that when the red pump laser interacts with it, a pair of entangled photons is emitted of the form

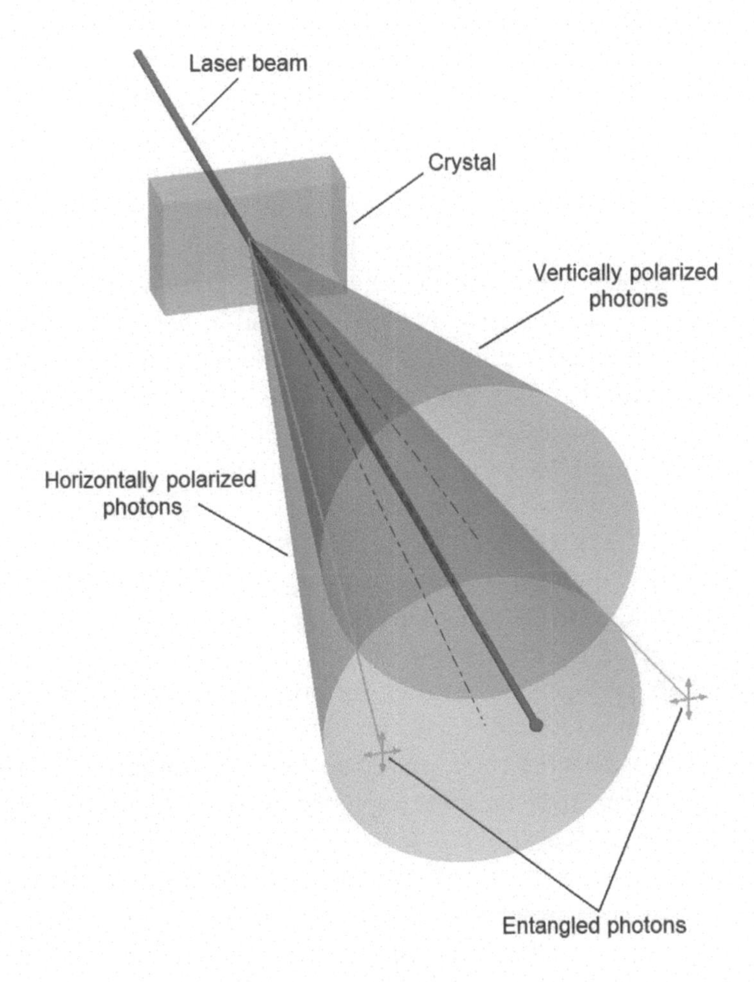

FIGURE 4.7 A laser beam pump (red) interacts with the optical crystal so that each laser photon has a small probability of downconverting into two daughter photons (green). All the daughter photons on the upper cone are V-polarized, and all those on the lower cone are H-polarized. If you place two pinholes at the two points marked with a cross, then what comes out is the entangled photon pair $|H\rangle_{\text{Alice}}|V\rangle_{\text{Bob}} + |V\rangle_{\text{Alice}}|H\rangle_{\text{Bob}}$.

$|H\rangle_{\text{Alice}}|V\rangle_{\text{Bob}} + |V\rangle_{\text{Alice}}|H\rangle_{\text{Bob}}$. Due to the double cone, it is easy to send one half of the pair to Alice and the other half to Bob. Given the issues with two-photon quantum-dot guns, this type of two-photon gun has been used and perfected for 30 years. The drawback is virtually all of the pump photons go through without converting into two daughter photons. So you have to start with a lot of the laser beam pump photons. When a laser beam photon does convert into two daughter photons, it does so at random. That is, you cannot predict when a pair will emerge, but rather estimate how many pairs are emerging per second. This is the set up for the entangled photon source on the Chinese satellite Mozi, probably one of the best in the world. The gizmo emits, on average, a million entangled pairs of photons per second. If you were to divide that second into a million time bins of one microsecond each (one microsecond is a millionth of a second), you would not see exactly one pair in each time bin. Some would have zero pairs, others would have one pair, others would have two pairs, and so forth. So this entangled photon is more like a shotgun or a machine gun than a pistol that fires one pair each time you pull the trigger, the latter of which would be a desirable double-photon gun. There is another issue. Sometimes two laser pump photons are converted into two daughter photons. Then instead of an entangled pair you randomly (and with a much lower probability), you get an entangled quadruplet. If you crank up the laser pump, you will then start to get entangled sextuplets (three photons in one direction and three in the other). This can be undesirable if you really just need one pair. However, this property can be used for some applications. If you put a photon-number-resolving detector in one of the outputs, and it clicks saying it detected three photons, then the number of photons in the other arm also collapses to three, even if the other arm is lightyears away.

Typically, what we do is run the machine continuously and do the cleanup in the data processing. For example, one of the first experiments carried out by the Chinese quantum satellite was to run the Bell test by a record-setting distance of 1,200 km. The entangled-photon source on the satellite looks precisely like that in Figure 4.7. The machine is firing photon pairs at random but quite a lot of them with an average of a million per second. (Only a few pairs per second make it to the ground due to photon loss in the atmosphere.) They can time-bin the pairs by pulsing the pump laser so that instead of a continuous wave it fires an intense laser pulse every nanosecond at the magic crystal. The bins are a nanosecond long. Since the device is spewing out a million pairs per second, and a nanosecond is a billionth of a second, only about one in a thousand of the time bins have a pair in them. The probability of having two pairs is minuscule.[12]

As you can see in Figure 4.7, the photon pair (indicated by the two green crosses) is separated, with one half going to a telescope controlled by Alice and a second to a telescope controlled by Bob. In China, the distance between two of these telescopes is around 1,200 km. The primary loss mechanism is not the absorption of the photons by the atmosphere but rather because the photon beams grow wider and wider as they approach the ground. As the light spreads out, most of the photons miss the telescopes entirely. Out of the million pairs per second generated, they only detect about 100 pairs per second at the ground.[13] Only about one in ten million of the nanosecond time bins have a pair in them. You have to wait a long time to collect data. If Alice gets a click at the detector at her telescope, say in time bin 137, and Bob gets a click at the detector at his telescope, also in time bin 137, then they are sure that the two photons came from one entangled pair. The probability of getting photons from different entangled pairs is negligible. On these few doublets that they do get, they now run the Bell test machinery and show the quantum theory is right to incredible precision over a record-setting distance. (The previous record was about 144 km, held by an Austrian group led by Anton Zeilinger.[14]) The Chinese group has just recently demonstrated a type of unbreakable cryptographic protocol. They were also able to teleport the state of one photon at Alice's ground station to Bob's ground station, a distance of 1,200 km. All new records! (See Figure 4.8.) We will discuss these protocols in more detail in the upcoming chapters.

---

12 Yin, Juan, et al. "Satellite-Based Entanglement Distribution Over 1200 Kilometers." *Science* 356 (2007): 1180–1184. Reprint: http://science.sciencemag.org/content/356/6343/1140; Preprint: https://arxiv.org/pdf/1707.01339.pdf.

13 A discussion of quantum experiments in Space is here https://en.wikipedia.org/wiki/Quantum_Experiments_at_Space_Scale.

14 A graphic of the teleportation experiment between the two Canary Islands may be found here, www.researchgate.net/figure/Experimental-setup-The-Bell-experiment-was-carried-out-between-the-islands-of-La-Palma_fig1_47644946.

FIGURE 4.8 This image is a time-lapsed photo of the quantum satellite Mozi passing over the Shanghai ground station that is on the roof of the University of Science and Technology of China and right above my office there. The green streak in the sky is the tracking and timing laser coming down from the satellite. The red beam is the tracking and timing laser from the ground station telescope going up to the satellite. The entangled photons are too dim to see. The tracking beams are spread out due to the time-lapse, which is why the fidgeting students are blurry. The tracking lasers are used to align the telescope on the satellite with that on the ground. They are also pulsed in nanosecond time bins for keeping track of each photon in a single entangled pair. In honor of me, this is called the Dowling Downlink.[15]

I SPY, WITH MY LITTLE EYE

To complement the discussion of photon emitters, we must now discuss the state of the art of the photon detectors, which are critical components of the quantum internet infrastructure. As far as photon detectors go, the human eyeball is a pretty good one. Experiments conducted in

15 The photo was taken by Weiyang Wang, who has given me permission to use the photo unaltered in this book, so long as I credit him and the University of Science and Technology of China, which I hereby so do.

2016 showed conclusively that the human eye could detect a single photon.[16] The first human-made photon detector (photodetector) was the Geiger counter.[17] (See Figure 4.9.) This device works well at detecting very short-wavelength photons called gamma-rays.

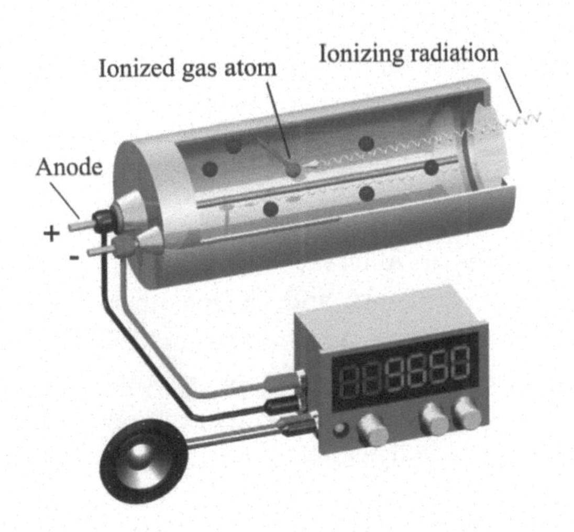

FIGURE 4.9 A Geiger counter consists of a metal tube with a metal rod running down the center of the tube. The tube is connected to one terminal of an electric power supply, and the rod (anode) is connected to the other terminal. There is a significant voltage difference between the tube and the rod. The tube contains atoms in a gas. When the gamma-ray (yellow squiggly arrow) enters the tube from the glass panel on the right, it hits one of the atoms and breaks it apart into an electron and a positively charged nucleus (a process called ionization). The charged particles then accelerate toward the walls of the tube, due to the high voltage, and begin smashing into and ionizing more atoms in a chain reaction. This reaction causes a voltage spike that can be read out on the detector and even heard as a click on the speaker.[18]

---

16 The experiment showing that people can see single photons was carried out by physicist Paul Kwiat – an endearing, bird-like, little man. It is described here www.nature.com/news/people-can-sense-single-photons-1.20282.

17 A description of Geiger's famous counter is here https://en.wikipedia.org/wiki/Geiger_counter.

18 The author of the figure grants permission to copy, distribute, and/or modify this document under the terms of the GNU Free Documentation License, Version 1.2 or any later version published by the Free Software Foundation; with no Invariant Sections, no Front-Cover Texts, and no Back-Cover Texts. A copy of the license is included in the section entitled GNU Free Documentation License: https://commons.wikimedia.org/wiki/File:Geiger-Muller-counter-en.png.

Hans Geiger was a German physicist working in the English laboratory of the New Zealand physicist Ernest Rutherford.[19] In that lab, he carried out an experiment where he fired helium nuclei at sheets of gold foil. This experiment provided evidence that atoms had a massive central, positively charged core. Rutherford worked out a theory that agreed with Geiger's data, and then Rutherford (but not Geiger) won the Nobel Prize in physics.

To detect how these helium nuclei (called alpha particles) scattered off the gold foil, initially, Geiger used his eyeball. He would sit in a dark room and stare at a sheet of fluorescent material, which would glow faintly when an alpha particle hit it. His dark-adapted human eye could detect the glow. His job was to measure how many alpha particles came out of the gold film at different angles. The experiment required millions of such measurements, and Geiger quickly grew bored of the tedious work. As the lore goes, he invented the Geiger counter to do the counting for him so that he could focus on other things. The counter turned out to be very good at counting gamma-ray photons in addition to alpha particles. It is still widely used today as a counter of high-energy particles and gamma-rays.

For our purposes, however, the Geiger counter is not useful. It does not work for the low-energy infrared photons that are compatible with the internet, and it does not tell you how many photons you got, that is, it is not number resolving. Because of the chain reaction, it will fire the same way if one or more photons hit it.

For quantum internet technologies, we have a list of requirements we would like for a perfect photodetector to meet. It should be able to detect single photons at infrared or visible wavelengths. It should be able to count photons, that is, to distinguish zero, one, two, etc. The device should operate at room temperature and not need cryogenic cooling. The detector should be efficient – that is when a photon hits it there is a 99% chance the detector goes click. The detector should not click when no photon is there. These requirements may seem trivial, but for most detectors, they will fire when nothing hits them. We call these firings dark counts because they occur even when there is no light impinging upon the detector. Dark counts are unwanted background noise. They arise from thermal fluctuations in the detector – the hotter

---

19 The story of Geiger himself is found here https://en.wikipedia.org/wiki/Hans_Geiger.

the detector the worse they get. One way to get rid of dark counts is to cool the detector, but then you run into the room-temperature issue. The detectors should be fast. If I'm using one in a quantum cryptosystem, I want it to be able to detect billions of photons per second and keep track of each one, so that the rate that I am transmitting cryptographic key is very high. Finally, for imaging and other purposes, the detectors should be arrayed in a two-dimensional grid with each detector a pixel in the grid. The more the megapixels, the better.[20]

E. T. PHONE HOME

Since the 1990s, there has been a lot of work on developing good photodetectors that work at optical or infrared wavelengths. Many of these use similar principles as the Geiger counter. Something has a high voltage applied to it, and when one or more photons arrive, it goes click. We use the latest version of this technology in your smartphone. Your phone's camera is composed of millions of pixels, each containing a charge-coupled device (CCD) detector.[21] This array of CCDs replaces the light-absorbing atoms in the old photographic film and allows the photo to be taken and stored digitally. We call the entire photodetecting apparatus a CCD array, as shown in Figure 4.10.

The atoms of the photographic film are now replaced with millions of CCD pixels, with each pixel a single-photon detector. A million pixels is a megapixel. When you buy a new smartphone, the more megapixels you have, the better the camera. (More accurately, the more megapixels *per square centimeter*, the better the camera.) To duplicate the quality of the photographic film, you want to have as many pixels per square centimeter as the film had atoms per square centimeter. We are not quite there yet, but at a certain point, an ordinary user can't tell the difference in the quality of a photo taken with a CCD array camera and one with the paper film. (The purists claim they can.[22]) The CCD array has many advantages over film. The primary one is that it is very sensitive to light. It will detect

20 A blurb on the DARPA photon detector program is here, www.darpa.mil/news-events/2015-01-13.

21 Ever wonder what CCD stands for in a CCD camera? Wonder no more, https://en.wikipedia.org/wiki/Charge-coupled_device.

22 On the merits of film over digital photography, https://en.wikipedia.org/wiki/Digital_versus_film_photography.

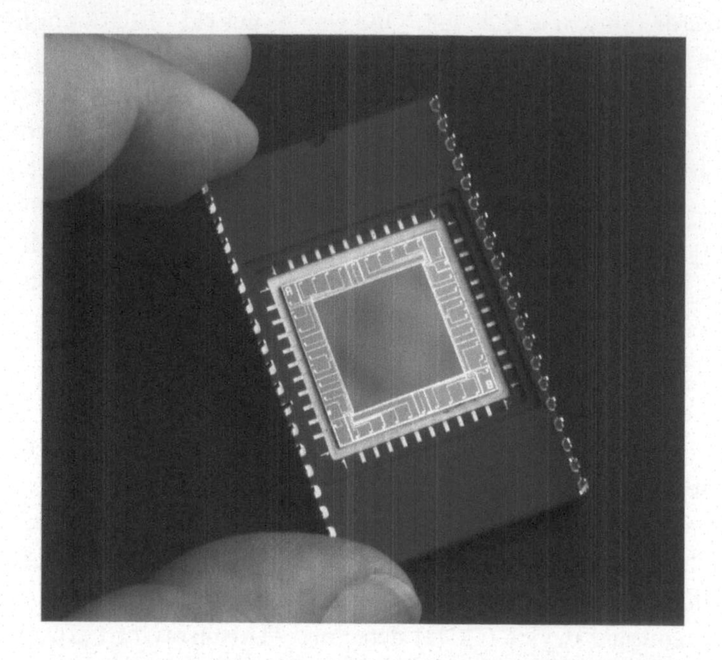

FIGURE 4.10 A CCD array like the one in your smartphone. The photo-taking region is the dull gray area in the middle that is composed of millions of CCD pixels.

about 95% of the light that falls on it. The CCDs perform better in low-light situations, and single photons are about as low-light as you can get.

The CCD camera's ability to work in low light means it requires much less overhead in design than a traditional film camera. The film does not collect light very efficiently so you either have to have very long exposure times or have lots of big lenses in the camera to direct more light onto the film. The CCD camera does not have this problem and can take very short exposures and the need for big lenses is moot. That's why they can fit the camera in a slim smartphone. The lens on the camera is just a speck of glass.

The CCD camera has the same drawback as the Geiger counter. Each pixel can count only either zero or one-or-more photons. The CCD process is similar in spirit to the Geiger counter process. A voltage is applied, so that when a photon arrives, it knocks an electron loose that then accelerates. That electron knocks another loose and then those two knock two more loose each, and you have exponential growth in the number of electrons in a runaway chain reaction. All those electrons excite a pixel. This exponential growth is good for photon efficiency, as the chain reaction amplifies even a feeble signal from a single photon, making it 95% likely that when a photon hits it, it will go click. However, due to that same chain reaction, the pixel is not photon-number resolving. If the dark counts are low, you seldom get a click if no photon is there, but you get the same click for one photon, or two photons, or three photons.

There is a trick to make the device number resolving. I call this a poor-person's-number-resolving detector, and we in our group have worked on them in collaboration with the team of Israeli physicist, Hagai Eisenberg, in Jerusalem.[23] The idea is that if you know you are only going to have a few photons arriving, as is the case for most quantum optical technologies, you can use the fact that the photon source is weak to do the following trick. Just the opposite of what you do in your smartphone, you yank the CCD array out of the phone and put a big fat lens in front of it that spreads out the incoming light beam. Let's say there are, on average, a hundred photons in the beam and million pixels in the CCD array. If the area of the beam is the same as the array, then each photon will inevitably hit a pixel and that pixel with fire with 95% probability. But the ratio of photons to pixels is a hundred by a million, which is one in ten thousand. Most of the pixels won't fire at all, and the odds of two photons hitting the same pixel are very small. In this way, you can ensure that the number of clicks is equal to the number of pixels. The CCD photon-number-counting detector has the advantage that it works well at room temperature and that it can count thousands of photons per second.

There is a photon-number-resolving detector that works really well, but you have to cool it to near absolute zero. It is based on superconducting technology, and we call it a transition-edge sensor or TES (pronounced Tess).[24] (See Figure 4.11.) It has a quantum efficiency of over 99%. That means it counts 99% of all the photons that hit it. For some quantum technologies, such as quantum sensing and imaging, this

---

23 Cohen, Lior, et al. "Absolute Calibration of Single-Photon and Multiplexed Photon-Number-Resolving Detectors." *Physical Review A* 98 (2018): Article Number: 013811. Reprint. https://doi.org/10.1103/PhysRevA.98.013811, Preprint: https://arxiv.org/abs/1711.03594.

24 The story of the transition-edge sensor is here https://en.wikipedia.org/wiki/Transition-edge_sensor.

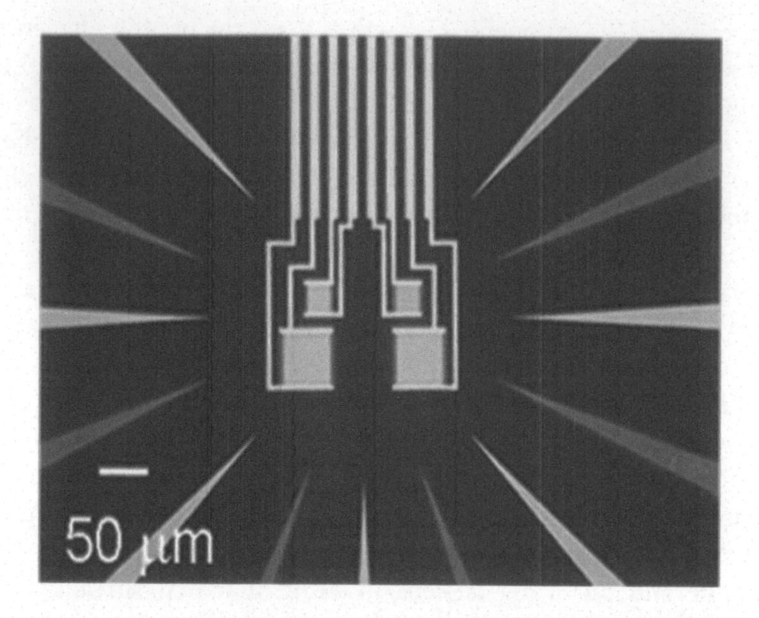

FIGURE 4.11 The TES is the orange and green square in the center that looks a bit like a spool of thread. It is attached to two gold electrodes relay the photon-number information off the chip.[25] The square is about 25 microns on a side.

type of efficiency is required. Due to the very cold operating temperature, the TES has an extremely low dark-count rate. When it clicks and tells you that you got two photons, you are very sure that two photons were there and not one or three.

The device is essentially a thermometer that can measure the heat deposited by even a few photons. Before any photons hit the square, it is superconducting, and so current flows through it unimpeded, which is the definition of superconducting. When a photon hits, it deposits a quantum of heat energy equal to the energy of the photon. This makes a hot spot on the square. Around the spot, the temperature rises a bit, so that near the spot, the metal of the detector is no longer superconducting. The slight change in the current is then picked up by the gold electrodes. If two photons hit, you get two such hot spots, and it tells you that two hit and so forth. The device is a bit sluggardly for

25 An article on the ability of a transition-edge sensor to resolve photon numbers. www.nist.gov/news-events/news/2011/11/adding-photons-tes.

quantum communications, but it has applications to quantum imaging and sensing, such as to laser-radar systems. It also works across a wide range of photon wavelengths, from microwaves to gamma-rays. Its main disadvantage is that it is slow. I can only count a few tens of photons per second. If the photons are carrying qubits of a quantum cryptographic key, this is a pokey communications system.

For some applications, such as a commonly used version of quantum cryptography, the number resolving feature is not required. The sender (Alice) is only sending one photon at a time, and so there is no need for the receiver (Bob) to resolve photon numbers. For these applications, a somewhat older technology, very similar to the Geiger counter, is used. These detectors are called single-photon avalanche photodiodes or SPADs.[26] (See Figure 4.12.) The avalanche is similar to the chain

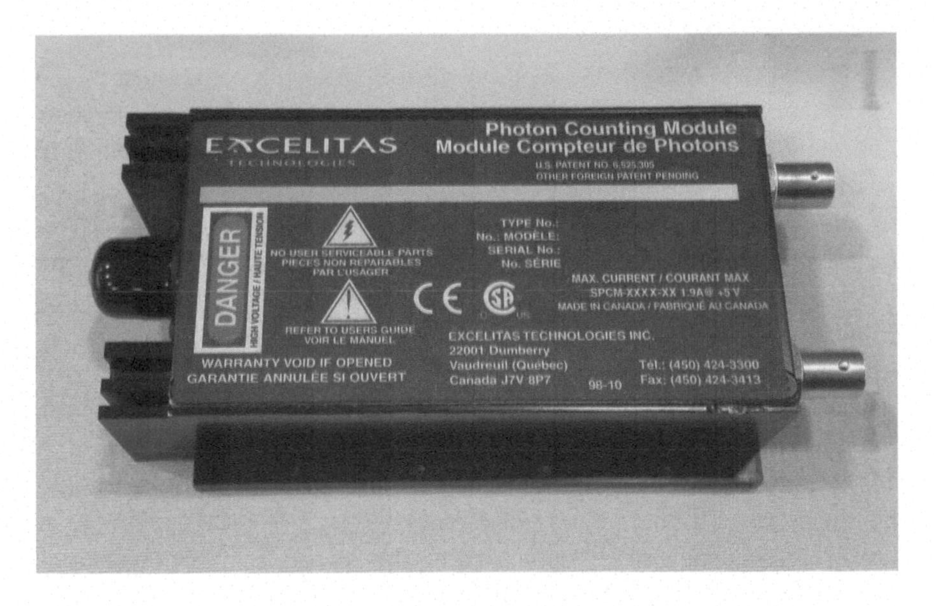

FIGURE 4.12 "Photograph of a single-photon counting module (SPCM) from Excelitas."[27]

---

26 Description of single-photon avalanche diode is here, https://en.wikipedia.org/wiki/Single-photon_avalanche_diode.

27 This figure is being reused under Attribution-ShareAlike 4.0 International license https://en.wikipedia.org/wiki/Single-photon_avalanche_diode#/media/File:Single-Photon-Counting-Module_Excelitas.jpg.

reaction in the Geiger counter. A diode is a device that conducts electrical current one way but not the other. The term single photon is a bit of a misnomer. SPADs (rhymes with cads) can only detect the difference between zero photons versus more than zero photons. Hence it is not number resolving. It gives the same click if one, two, three, or more photons hit, but it does not click if no photons hit. We sometimes call them on–off detectors rather than single-photon detectors for this reason. As discussed above, if Alice is only sending single photons, and Bob gets a click, he can somewhat safely assume that it was only one photon because that is all Alice sent. It also works at room temperature, but with a relatively high dark-count rate, since at this temperature, the thermal noise triggers fake-photon detection events – the dark counts. The dark counts can be reduced by cooling the device, making the device smaller, or by time-gating Alice's incoming photon so that Bob only has the SPAD turned on in a short time window when he expects a photon from her.

SPADs have an amusing history. Photon detector companies in the 1970s and 1980s noticed that every once in a while, one of their commercial avalanche photodiodes (APDs) had this magic ability to resolve zero or more photons. Something like one out of a thousand of their detectors was a SPAD. They had no idea why this was so, but they were able to take these magic detectors put them in a box labeled SPAD instead of APD and then jack up the price by thousands of dollars. Later, in the 2000s, people learned the trick of making good SPADS directly without this costly trial and error process and so the price dropped quite a bit. Unlike TES, the SPADS can count thousands or millions of photons per second and so it is more useful for high-speed quantum communication. I could go on and on about different detectors but suffice it to say they get better every year. But there's more to the quantum internet components than just the detectors. For many applications, they must be able to detect the polarization of a single photon and not only that the photon is there.

POLARIZE! LET NO PHOTON EVADE YOUR EYES!

There are a number of internet protocols that require that we detect the photon by projecting it onto a specific polarization state. These include quantum cryptography as well as things like anonymous broadcasting,

Bell tests, and so forth. We'll discuss these protocols in the next couple of chapters. I briefly alluded to polarization analyzers in chapter two but did not go into much detail about how you make them. Here, I explain this a bit more. In Figure 2.14, I sketched a polarization analyzer. Let's see how we do it in the lab. The trick is to use a polarizing beam splitter, as shown in Figure 4.13.[28]

In the figure, the polarizing beam splitter, made from Icelandic spar, as we have discussed in chapter one, separates the H from the V photons and routes them to the right or down where there sit two single-photon detectors (not shown). The air gap allows for a separation into right angles so the photons can easily be routed in different directions. If the right detector clicks, the photon collapses into V. If the down detector clicks, the photon collapses into H. You can also

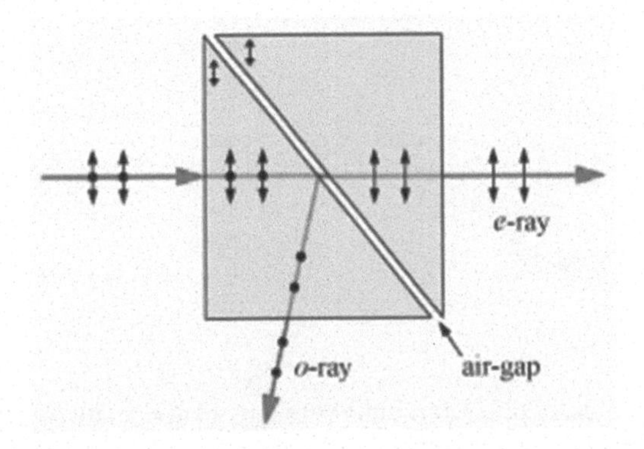

FIGURE 4.13 A polarizing beam splitter that separates vertically polarized photons extraordinary rays (e-rays) from horizontally polarized ordinary photons (o-rays). These terms are somewhat archaic. There are three polarization axes. Two, the ordinary, allow the light to propagate at the same speed, but the third, the extraordinary, enables the light to propagate at a different speed. That speed difference is what separates the two polarizations. The H photons then go to one detector down below, and the V photons go to another detector to the right.[29]

28 A description of a polarizing beam splitter https://en.wikipedia.org/wiki/Glan_prism.

29 The second kind of polarizing beam splitter https://en.wikipedia.org/wiki/Glan%E2%80%93Foucault_prism.

make the device separate ± 45°-polarized photons from each other.[30] Such a device is called a wave plate. (See Figure 4.14.) We now can integrate these onto computer chip-sized devices, and the detector can switch from an *H/V* detector to a ±45° detector very rapidly. For the Bell test, we need even weirder angles like 22.5°/67.5°, but the polarization rotators can handle any angles. In this way, the photon detectors, the polarizing beam splitter, and the polarization rotators can be put into a box labeled a polarization detector. This type of sensor is essential, not just for Bell tests, but also specific cryptographic protocols that we will discuss in the upcoming chapters.

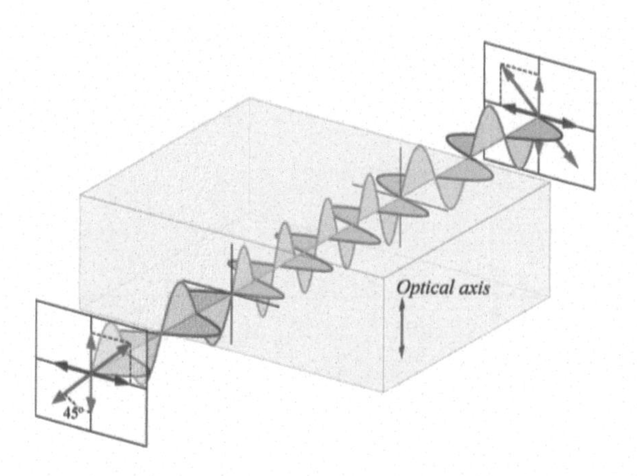

FIGURE 4.14 "Linearly polarized light entering a wave plate can be resolved into two waves, parallel (shown as green) and perpendicular (blue) to the optical axis of the wave plate. In the plate, the parallel wave propagates slightly slower than the perpendicular one. At the far side of the plate, the parallel wave is exactly half of a wavelength delayed relative to the perpendicular wave, and the resulting combination (red) is orthogonally polarized compared to its entrance state."[31] To rotate the polarization by 45°, we use what is called a quarter-wave plate, which is just half the length of the half-wave plate. Combining half- and quarter-wave plates at different angles allows us to get any rotation, such as the 22.5°/67.5° required for a Bell test.

30 Description of a polarization rotator https://en.wikipedia.org/wiki/Polarization_rotator.
31 How to make a wave plate from a rotator https://commons.wikimedia.org/wiki/File:Waveplate.png.

RARE AS A TWO-DOLLAR BELL

The polarization detector, discussed above, projects the polarization state of a *single* photon onto any polarization axis. For specific protocols such as teleportation and entanglement swapping, presented in the next chapter, we need a type of two-photon detector that, instead of projecting into a single-photon polarization basis, projects into a kind of two-photon polarization basis called the Bell basis. The Bell basis is given by the two-photon Bell states, discussed in chapter three. There are four two-photon entangled states, called the Bell states, and they are $|\Phi^{\pm}\rangle_{AB} = |H\rangle_A|H\rangle_B \pm |V\rangle_A|V\rangle_B$ and $|\Psi^{\pm}\rangle_{AB} = |H\rangle_A|V\rangle_B \pm |V\rangle_A|H\rangle_B$. Given any one of them, we can manipulate it into any other one using polarization rotators and wave plates. Here, the subscripts denote photon A for Alice and photon B for Bob.

For single photons, we are primarily interested in projecting them into *one of two* polarization basis states, either H/V or ±45°. For two entangled photons, we are mostly interested in projecting the joint state of the two photons into *one of four* of the Bell basis states, given above. A device that does this can carry out a Bell measurement.[32] If the detector works correctly, then the incoming two-photon state randomly collapses into one of the four Bell states upon measurement. Just like the single-photon measurement *gives one of two outcomes*, corresponding to say H/V, the Bell measurement *gives one of four outcomes*, corresponding to either the two projections $|\Phi^{\pm}\rangle_{AB}$ or the two projections $|\Psi^{\pm}\rangle_{AB}$. That is, the measurement outcome can be described by four classical numbers, 1, 2, 3, 4, which, when written in binary, are 00, 01, 10, 11. As we shall see in the next chapter, Alice needs to transmit one of these four numbers to Bob to complete the teleportation of a quantum state.

If one tries to carry out a Bell measurement using only beam splitters and polarization rotators, the measurement only succeeds 50% of the time. It destroys all information about the Bell state the other 50% of the time. For some applications, such as quantum key distribution, this failure rate goes into the photon-loss budget, and Alice and Bob simply repeat the experiment over and over until it works. Luckily, the measurement outcome tells you when it has worked and when it has

32 How to make a Bell measurement https://en.wikipedia.org/wiki/Bell_state#Bell_state_measurement.

not. For other applications, such as quantum teleportation, this 50% success probability is a problem. Captain Kirk would think twice about entering the transporter on the Star Ship Enterprise if the device successfully teleported him to the planet below only half the time but blew him into molecular bits the other half of the time. The same is true for a precious quantum state that might be the output of a quantum computer – a state which you could not afford to lose! Thankfully there are nontrivial ways to boost this success probability to 100%, so we never lose James Tiberius Kirk. The improved method involves nonlinear two-photon gates that we use in optical quantum computing. For those, the Bell measurement works 100% of the time. We have enough of the building blocks now to start describing different protocols that will run on the quantum internet, and we do that in the next chapter.

CHAPTER 5

The Second Quantum Revolution

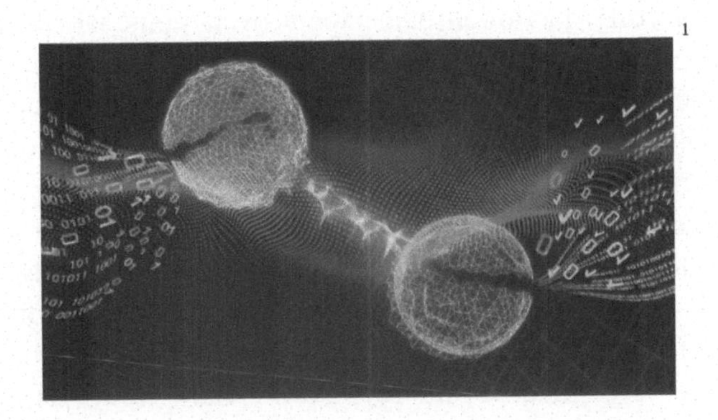
[1]

We are currently in the midst of a second quantum revolution. The first quantum revolution gave us new rules that govern physical reality. The second quantum revolution will take these rules and use them to develop new technologies.[2]

1 Permission to use this designed graph is given to me by the University of Science and Technology, Shanghai, China.

2 Dowling, Jonathan P. and Gerard J. Milburn. "Quantum Technology – The Second Quantum Revolution." 2003. https://doi.org/10.1098/rsta.2003.1227; https://arxiv.org/abs/quant-ph/0206091. The title of this paper, suggested by Milburn, was to be only "Quantum Technology". But I, being a lover of compound titles, decided to add "The Second Quantum Revolution". Before 2002 this latter phrase turns up only once in Google in reference to the change between the old Bohr model and the modern quantum theory (circa 1925). If I Google this exact phrase right now – I get nearly 17,000 hits. There is even a book – not written by me! – titled "The Second Quantum Revolution" that does not even cite me! (But I'm not bitter.) I should have trademarked the term when I had the chance! (Upon writing this, I double-checked, and it is not trademarked by anybody. I just filed for one.)

WHAT IS QUANTUM TECHNOLOGY?

As I have mentioned, the first paper to clearly define quantum technology was my 2003 paper with Gerard Milburn, "Quantum Technology: The Second Quantum Revolution". This paper started out as something I was writing for a NASA technical brief. At the time, in the early 2000s, there was quite a bit of funding floating around in NASA for nanotechnology. When asked by NASA managers, "What is quantum technology – is it just *smaller* than nanotechnology?" I replied, "No – it is much *weird*er than nanotechnology!" To expand on this notion, I wrote up my thoughts as this NASA brief. Around the same time, Gerard Milburn was asked to write a review article *for the Proceedings of the Royal Academy of London*. This is a very prestigious journal which, for hundreds of years, has published papers by famous scientists such as Isaac Newton (1671),[3] James Clerk Maxwell (1880),[4] Michael Faraday (1843),[5] Subrahmanyan Chandrasekhar (1929),[6] and Stephen Hawking (1965).[7] This was the first time (and probably the last) a lowly NASA technical brief ended up in this journal. Milburn and I polished the thing up and added nice photos, which we borrowed from the internet, and submitted it. It was so well received that one of our figures made the front cover of the journal, even though the journal forgot to get permission from the electron photographer to use it. If you Google the exact phrase "second quantum revolution" right now, you get about 17,000 hits. If you restrict the date to before when our paper first appeared on the internet, in 2001, you get nearly zero hits. We should have trademarked this phrase. In our paper, Milburn and I did not give a very strict definition of what quantum technology is. We simply discussed things we thought should be quantum technology and left out other things that we thought should not. However, now, as the field is taking off, I'd like to put down here my definition of what it is.

3 Newton, Isaac. "A Letter of Mr. Isaac Newton Containing His New Theory about Light and Colors." 1671. https://doi.org/10.1098/rstl.1671.0072.

4 Maxwell, James Clerk. "On a Possible Mode of Detecting a Motion of the Solar System through the Luminiferous Ether." 1879. https://doi.org/10.1098/rspl.1879.0093.

5 Faraday, Michael. "Experimental Researches in Electricity." 1843. https://doi.org/10.1098/rspl.1837.0010.

6 Chandrasekhar, Subrahmanyan. "The Compton Scattering and the New Statistics." 1929. https://doi.org/10.1098/rspa.1929.0163.

7 Hawking, Stephen William. "On the Hoyle-Narlikar Theory of Gravitation." 1965. https://doi.org/10.1098/rspa.1965.0146.

To be called quantum technology, a technology must exploit at least *one* of the three features that Einstein disliked about quantum theory:

1. quantum uncertainty,
2. quantum unreality,
3. quantum nonlocality.

All of these features I have highlighted in previous chapters. It is in the last of these three requirements that I am being very careful not to use the word entanglement – as there has sprung up recently in the literature a new field that people are calling "classical entanglement". This is a completely dishonest and deliberate misuse of the word entanglement. The term is promulgated by people – who are not working on quantum technology – in order to be able to claim that they are. They then hope the funding agencies don't notice the difference.[8] The same kind of chicanery occurred 20 years ago when the government-funding sluice gates opened up with rivers of money earmarked for nanotechnology. Wanton chemists would claim they were working on nanotechnology because – in their vials of goop – the molecules were nanometers across. Libeling laser jocks promoted their work as nanotechnology because the wavelength of the light in their laser pointers was hundreds of nanometers long. To preempt this kind of dishonesty from recurring again, we need a concrete and simple definition of what quantum technology is, and I have just given you one. For the rest of this book, when I write entanglement, I mean only the type that encodes nonlocality. I can now roll out things you might want to have for your quantum internet, and I'll explain why each one of them is a quantum technology by the above standard. The gold standard is that the technology meets all three – and not just one – of the above requirements. Right now, thanks to the new U.S. National Quantum Initiative, we have people with money – who know nothing about quantum technology – funding people without money – who also know nothing about quantum technology.

The rest of this chapter will highlight quantum technology protocols or devices that will be the critical core components of the future quantum network. A good place to start is with a discussion of quantum

8 Karimi, Ebrahim, and Robert W. Boyd. "Classical Entanglement?" 2015. https://science.sciencemag.org/content/350/6265/1172.

random-number generators (QRNGs) – no quantum internet is complete without one!

QUANTUM RANDOM-NUMBER GENERATORS

In the last chapter, we discussed the building blocks of the quantum internet. In this chapter, we shall show how to start hooking these together into useful network protocols. I was unsure where to put the QRNG. Due to inaction on my part, they have defaulted into this chapter. I think that choice was perhaps because they are a bit complicated and themselves composed of the building blocks in the last chapter. Also, some QRNGs can span the quantum network and thus generate secret codes. And finally, no network can function securely without a good source of random numbers. We'll cover several types of random-number generators.

The photon-based QRNG of the first kind we discussed in chapter two. (See Figure 2.4.) A single-photon gun shoots single photons – what else? – at a 50–50 beam splitter. For each photon, we can write the quantum state after the beam splitter as $|1\rangle_C|0\rangle_D + |0\rangle_C|1\rangle_D$, where the notation indicates that the photon is in a cat-like superposition state of the photon going up to detector C, with nothing going to detector D, and the photon going right to detector D, with nothing going up to detector C. Written this way we can see that the state of the photon is *unreal*, it is not either at C or D until one of the detectors clicks, and thus the machine satisfies the second criterion for a quantum technology. Additionally, if the beam splitter is a perfect 50–50 splitter, then it is utterly uncertain which detector will click for any photon. All we can say is that 50% of the time detector C clicks – and 50% of the time detector D clicks – but which will click is unpredictable by any means. Hence the device satisfies criteria one – *uncertainty*. Finally, the photon is in an entangled state with nothing. More correctly, the state $|0\rangle$ is called the vacuum state. In quantum theory, the vacuum is not empty but instead contains zero photons, where zero photons mean something.[9] The state after the beam splitter is an entangled state of one and zero photons, and hence it satisfies the third criterion – nonlocality.

To get a string of random numbers out of this, we'll work in the computer binary-bit basis of classical 1s and 0s, which are different from the 1s and 0s of the state. If detector C clicks, we assign that

---

9 See https://en.wikipedia.org/wiki/Quantum_fluctuation.

event a 1. If detector D clicks, we assign that a 0. As the machine plugs away, we eventually establish a long random string of classical 1s and 0s such as

001001000011111101101010100010001000010110100011000010001101001 . . . .

The next question you might have is, what do we need random numbers for? Interestingly, the most lucrative commercial use of such things is in the gambling industry. Random numbers are used to generate the random outputs of a slot machine or keno. If a player could surreptitiously skew the random numbers in the machine, then the player could change the odds to improve their chances of winning. That can be deftly done with a classical random-number generator, but it is impossible to do with a QRNG, due to uncertainty, unreality, and entanglement.[10] Verifiably random numbers are also crucial in whopping computer simulations that use a type of sampling called the Monte Carlo method.[11] If the numbers are not random, the computer simulation will produce nonsense.[12] This book is not about the gaming industry – random numbers are also useful in quantum cryptography – as we will discuss in the next chapter.

The photon-QRNG of the first kind has been developed commercially, but it has taken some years to get the bugs out.[13] If the beam splitter is not 50–50, for example, then the random numbers will be skewed. We have dealt with most of these bugs, but still, we have no reliable test to prove that these QRNGs of the first kind give genuinely random numbers. Thus, quantum technologists have introduced QRNGs of the second kind that exploit a nonlocality test to show the numbers are honest-to-God random.

---

10 "Quantum Random Properties." https://en.wikipedia.org/wiki/Hardware_random_number_generator#Quantum_random_properties.

11 "Monte Carlo Method." https://en.wikipedia.org/wiki/Monte_Carlo_method.

12 In one apocryphal story, some astrophysicists were using a bad "pseudo" random number generator in their simulations of the structure of the Universe. The simulation showed that the Universe should contain giant superclusters of galaxies that form in large, periodic, wall-like structures that span the entire Universe. Astronomers spent months of valuable telescope observing time looking for these predicted structures without finding anything like them. In the end, the structures were a result of using a bad random number in the computer simulation – they did not exist.

13 More on the IDQ random-number generator can be found here www.idquantique.com/random-number-generation/products/quantis-random-number-generator.

QRNGs of the second kind exploit all three properties that define quantum technology, quantum uncertainty, quantum unreality, and quantum nonlocality – that is, in addition to the first two, they exploit the quantum special sauce, which is entanglement. This device is identical to the entangled source used in the Ekert 1991 quantum cryptography scheme.[14] As described in chapter three, the state emitted by the central source is $|H\rangle_{\text{Alice}}|V\rangle_{\text{Bob}} + |V\rangle_{\text{Alice}}|H\rangle_{\text{Bob}} = |+\rangle_{\text{Alice}}|-\rangle_{\text{Bob}} - |-\rangle_{\text{Alice}}|+\rangle_{\text{Bob}}$. That is, it is entangled in both the H–V basis and the ±45° basis. First, we recall how we make random numbers. If Alice measures her photon in the H–V basis and gets a V, then Bob must get an H. They agree to call this outcome 1. Alternatively, if Alice obtains an H, then Bob must see a V. They agree to call this outcome 0. The randomness of these numbers is guaranteed by the first two ingredients of quantum technology – uncertainty and unreality. However, recall that there could be undetectable gremlins infesting the source and the detectors – fiddling with those sensors in such a way that the random numbers are not really random at all! They could be skewed so that the gremlins could routinely beat the casino at slot machines. As discussed in great detail in my previous book, in the section on loopholes, you can rule out gremlins in the detectors so long as the probability that the detector detects a photon is higher than about 82%. NIST has such detectors in abundance – hence they can rule gremlins in the detectors out. Next – to take care of any diabolical imps in the source, we periodically run the Bell test by flipping the polarization analyzers into lots of different angles and then feeding the data into a computer that spits out a number. If the number is even a little bit bigger than two, the machine passes the test. If the number is less than two, the gadget fails. It will fail if gremlins are, say, sending out two photons that are not real-McCoy entangled but have classical correlations to make them look as if they are entangled. In this way, the gremlins could skew things up. This Bell-test setup is almost perfect. If we put the source and detectors in a box with a computer chip, the box should spew out *bona fide* random numbers – unpredictable by any means. (See Figure 5.1.)

14 See, https://en.wikipedia.org/wiki/Quantum_key_distribution#E91_protocol:_Artur_Ekert_.281991.29, see also, Dowling, Jonathan P. *Schrödinger's Killer App: Race to Build the First Quantum Computer*. Taylor & Francis Press, 2003, Ch. 5.

FIGURE 5.1 The entangled-photon source at the center of the NIST quantum random-number generator. "The method generates digital bits (1s and 0s) with photons or particles of light. An intense laser hits a special crystal that converts laser light into pairs of photons that are entangled, a quantum phenomenon that links their properties. These photons are then measured to produce a string of genuinely random numbers." The orange light is from the pump laser that pumps a nonlinear crystal, which is glowing in the center. When an orange pump photon downconverts in the crystal, it produces two daughter photons, which are entangled. The entangled photon pairs have a wavelength in the infrared, and cannot be seen with the naked eye.[15]

There is one more loophole – the locality loophole. If the source and detectors are very close together, it is possible that a dearly departed gremlin from the distant past mucked around with the source and detectors at the factory that built the gadget. That mucking is then in the hardware from the beginning, and we are again all skewed. Alternatively,

15 Photo credit to my old pal, Dr. Krister Shalm, who spends *way* too much time reading fortunes from fortune cookies he gets from a Chinse restaurant near NIST where he has lunch *every single day*, www.nist.gov/news-events/news/2018/04/nists-new-quantum-method-generates-really-random-numbers. This image is the property of the U.S. Government and not subject to copyright.

there could be gremlins sitting on the source and detectors talking on their cell phones and relaying the measured outcomes to each other so that their pals can skew things up. The solution to this is not particularly convenient, but it does the job. The idea is illustrated in Figure 3.8. Now, in addition to everything we have done so far, we place the detectors far from each other and the source – so far that if gremlins are sitting on the stations, they would have to communicate faster than the speed of light – which is impossible. In addition to the Bell test, we also use the property of the state $|H\rangle_{\text{Alice}}|V\rangle_{\text{Bob}} - |V\rangle_{\text{Alice}}|H\rangle_{\text{Bob}} = |+\rangle_{\text{Alice}}|-\rangle_{\text{Bob}} - |-\rangle_{\text{Alice}}|+\rangle_{\text{Bob}}$ that it is maximally entangled in both H–V or ±45° – at the same time! That is quantum unreality at work. Alice and Bob then can randomly switch between H–V measurements and ±45° measurements – or any other angles. If all is done sufficiently quickly, then the gremlins cannot interfere with the system unless they have faster-than-light communicators. With all these checks and balances in place, the numbers are certifiably random by the laws of quantum theory and Einstein's theory of relativity. One or the other, or both of these theories, would have to be wrong for this scheme to fail.

In summary, there are QRNGs capable of producing authentic random numbers that can be used by the gambling industry, cosmologists, and cryptographers. The numbers are certifiably random and certifiably tamper-proof. We will talk about the cryptography applications of these next.

CRYPTOGRAPHY

There is an entire chapter on this topic in my previous book. I'll skim the details of point-to-point quantum crypto and then lay out how this can be installed in a quantum network. There are quite a number of quantum cryptography protocols, but I will focus on the two in everyday use today.[16] The first is BB84, invented by Bennett and Brassard in 1984 (hence the acronym). It uses the first two ingredients of quantum technology but not quantum entanglement. The second is the E91 protocol invented by Ekert in 1991. The Ekert scheme explicitly exploits entanglement and thus meets all three criteria for quantum technology. Why do we need quantum cryptography? The term is a bit of a misnomer. The real name for it is quantum key distribution (QKD). Once the two parties, Alice and Bob, share a long list of random numbers, then they use that key to transmit messages over an open channel such as Twitter, Instagram, Facebook, or

16 "Quantum Key Distribution." https://en.wikipedia.org/wiki/Quantum_key_distribution.

WhatsApp. The key is guaranteed by quantum, and the security of the key guarantees the messages. While the term quantum cryptography is widely used, the correct term is quantum key distribution (QKD). It is the secret key that is transmitted quantumly. The encryption then uses this classical key to send secret messages. The former is key distribution, and the latter is cryptography – and only the key delivery is quantum. First, let us discuss the most widely used of classical crypto schemes in order to see how the quantum version is better.

CLASSICAL CRYPTOGRAPHY – RSA PUBLIC-KEY ENCRYPTION

Classically, the two most used cryptographic schemes are public-key encryption and private-key encryption. The most widely used of the former is the RSA public key. The RSA public key was invented in 1997 by Rivest, Shamir, and Adleman (RSA).[17] Alice picks two large prime numbers,[18] p_A and q_A, at random. She then multiplies them together to get a composite number, $p_A \times q_A = c_A$. In Catholic school, I learned from the nuns that multiplication is much easier than division. That observation holds even more so if p_A and q_A are humongous numbers. The public-key protocol then lets Alice publish her composite number c_A on her web page. Anybody, particularly Bob, can use this public key c_A to send Alice an encrypted message. However, once Bob encrypts his missive, even he cannot decrypt it because that requires that he knows the two private keys p_A and q_A, which Alice keeps secure.

A handy analogy is to consider a lockbox with two locks on it. One lock corresponds to the public key c_A, and the other lock corresponds to the private keys p_A and q_A. Alice sends the box to Bob. He puts his message in it, using the public key c_A, and then locks it back up. At this stage – he cannot open it again – only Alice can open the box with her private keys p_A and q_A. Bob ships the box back to Alice; she opens it and decrypts the message using her private key. Of course, this analogy of the box is not what is done in the real world – nobody is shipping boxes – the operations are all done with zeros and ones on the internet via computer programs at Alice and Bob's ends. The computer algorithm that encrypts Bob's message

17 For more on public-key classical cryptography, see https://en.wikipedia.org/wiki/RSA_(cryptosystem).

18 More on prime and composite numbers can be found here https://en.wikipedia.org/wiki/Prime_number.

is called a trapdoor function. Bob opens the trap door with Alice's public key and drops in his message. Then the trapdoor slams shut – Bob can't get his message back.

Let's use a simple example to see how this all works. To choose her random prime numbers, Alice uses a random prime-number generator on her computer or on a secure site on the web.[19] For example, she gets,

$$
\begin{aligned}
p_A &= 104,955,930,254,087,466,380,696,646,332,770,\\
&\quad 990,366,887,256,024,656,457,793,486,151,094,\\
&\quad 227,818,938,697\\
q_A &= 10,454,770,931,944,725,031,480,481,057,839,\\
&\quad 728,022,209,464,327,264,305,980,736,084,346,\\
&\quad 790,983,789,849\\
p_A \times q_A = c_A &= 1,097,290,208,755,651,581,961,703,033,557,\\
&\quad 293,659,924,605,457,339,441,941,744,078,469,\\
&\quad 544,270,081,454,625,655,423,653,707,245,932,\\
&\quad 137,113,553,270,864,671,956,751,769,326,964,\\
&\quad 765,753,869,497,930,061,886,753
\end{aligned}
$$

This is called a 256-bit public key because it takes 256-bits (0s and 1s) to represent the composite number c_A in binary notation. For example, the popular messaging application WhatsApp uses 256-bit encryption.[20] Now we remember the words of the nuns from Catholic school – multiplication is easy, but division is hard! In order for Bob to get his own message back, or worse, for Eve – the eavesdropper – to intercept Bob's message, Bob or Eve would have to take the insanely long composite number c_A and divide it out (factor it) back into the two primes that Alice is hoarding. To do this on a classical computer would take longer than the lifetime of the universe, and thus computer scientists thought this public key was uncrackable.

19 Here is an online random prime-number generator for you to play with https://asecuritysite.com/encryption/random3?val=16.

20 The encryption on *WhatsApp* is so good that a number of countries ban it because these governments want to see what their citizens are up to. Since WhatsApp is now owned by Facebook, it is safe to assume that Facebook and the U.S. Government have put a back door into WhatsApp that allows the government to read the encrypted communications. By law, the U.S. is prohibited from spying on its own citizens, but there is a concern that foreign terrorists could be conspiring on WhatsApp to plan attacks. More on the app and its history is here https://sco.wikipedia.org/wiki/WhatsApp.

For Alice to send a secret message to Bob, Bob must first establish his own private pair of prime numbers and then multiply them together and post the result on his web page. Then Alice uses Bob's composite number to encrypt messages to him, and Bob decrypts them with his private key. As long as we cannot factor the public key into the private keys, this protocol allows for any two parties to communicate in secret. Once you encrypt the message, you can send via the mobile phone system, the internet, a telegraph, or smoke signals. It is this public-key encryption that secures all your financial transactions, medical data, Air Traffic Control, the power grid, and WhatsApp communiques. The scheme is used nearly universally by industry, the financial sector, and for most government communications. The world has become utterly and nearly irreversibly reliant on the security of public-key encryption. If you found a computer algorithm to break it, and posted that algorithm on the internet, you would cause the entire economy of the world to collapse in days. I illustrate the entire scheme in Figure 5.2.

FIGURE 5.2 Public-key cryptography – creating a key pair. Two big random numbers are used to create a key pair – the public key and the private key. Without knowledge of the random numbers, it was thought to be impossible to recreate the private key from the public key. The public key can be freely published online.[21] The sender Bob (on the left) uses Alice's public key – the product of two large prime numbers – to encrypt the plaintext into the encrypted cyphertext. Only the receiver Alice, who holds the two prime numbers in secret, can decrypt the cyphertext.

21 Image licensed under an agreement between Taylor & Francis Publishers and Shutterstock. www.shutterstock.com/image-illustration/public-key-encryption-224607571.

Unfortunately for all of us, there is now a sure way to crack this key – and it is with a quantum computer. However, nobody has ever proved that the public key can't be broken – even on a classical computer![22] The computer scientists told everybody, "We think we cannot crack this code by any means – but we're not sure! Trust us anyway?" Our faith in the prognostications of computer scientists has now put the entire world in a very precarious position. Quantum computers – capable of hacking the internet – will be ready in as little as ten years.[23] Even worse, eavesdroppers all over the world are now vacuuming up all your encrypted data and then storing it to decrypt later when they get their hands on a quantum computer. There are data like medical records, Swiss bank accounts, or the plans for building an atomic bomb that people want to keep secure for decades or centuries. They are all using public-key encryption to secure that stuff – and they are all hosed. There is another classical cryptography system, which is far less used, called the one-time pad.

CLASSICAL CRYPTOGRAPHY – ONE-TIME-PAD PRIVATE-KEY ENCRYPTION

There is another popular – but much less widely used – private-key encryption scheme called the one-time pad. The name comes about because – in the old days – you encrypted your message using a page from a pad of paper, as illustrated in Figure 5.3 That is an old one-time pad developed by the U.S. National Security Agency (NSA).[24] The one-time pad encryption scheme, invented in the 1880s, was proven *unbreakable by any means* by the father of classical information theory, American mathematician and engineer, Claude Shannon.[25] To see how it works, let's look at the pad in Figure 5.3. The first step is that Alice and Bob must each have a copy of the same pad – and nobody else should have one. Anybody with a copy could read the encrypted messages. A bigly drawback

22 In 1992, two years before Peter Shor announced his quantum factoring algorithm, the film *Sneakers* appeared. It starred everybody from Robert Redford to James Earl Jones. It is a thriller whose improbable plot is that a mathematician discovers a fast, classical-factoring algorithm, which he then programs into a telephone answering machine. The mathematician was murdered by the bad guys who then stole the answering machine. The rest of the movie involves the good guys trying to steal it back. See https://en.wikipedia.org/wiki/Sneakers_(1992_film).

23 This ten-year estimate was given by Astrid Elbe, Intel Labs Europe, at a conference I attended in July of 2018. Details of the conference can be found here www.muenchner-kreis.de/veranstaltungen/seit-2010/quantum-technology.html.

24 For more on the one-time pad, see https://en.wikipedia.org/wiki/One-time_pad.

25 The life and times of Claude Shannon can be found here https://en.wikipedia.org/wiki/Claude_Shannon.

of the pads is that we must distribute them securely so that nobody can copy them in transit.

After Alice and Bob have their identical copies of the pad, they set up a protocol that at midnight on day one – they will use the first page of the pad and on day two the second and so on. The encryption protocol is in the table on the right, where you can see in each row the letters of the alphabet in alphabetical order at the top of the row and all the letters of the alphabet in random order at the bottom of the row. The ordering of the random letters must be *truly* random.

Let us suppose that Alice decides to send the message SCHRODINGERS CATS WERE ALIVE to Bob. The agreement is that at midnight they begin with the first row at the top labeled A in the figure. The encryption goes as follows: take each letter in the message and find that letter in the top of row A and then replace it with the corresponding random letters in the bottom of row A. The transcription goes from top to bottom.

S	C	H	R	O		D	I	N	G	E		R	S	C	A	T		S	W	E	R	E		A	L	I	V	E
H	X	S	I	L		W	R	M	T	V		I	H	X	Z	G		H	D	V	I	V		Z	O	R	E	V

We have broken up the message into five chunks of five letters each, as indicated on the left of Figure 5.3.[26] The rule then is to use one line of 25 letters per one row of the encryption key and never to use the same row twice. Now we make an essential *point.* Suppose Bob tells Alice, over an unsecured open channel such as email or a postcard, that the message did not go through and that she should send it again. (As long as Bob reveals nothing about the pad, this communication provides no information to an eavesdropper.) Alice must *never* retransmit HXSIL WRMTV IHXZG HDVIV ZOREV ever again! Violating this rule is called *reusing the pad – and reusing* the pad violates one of the assumptions that go into Shannon's unbreakablity proof.

In the lingo of crypto, the message before encryption, SCHRODINGERS CATS WERE ALIVE, is called the plaintext and the message after encryption, HXSIL WRMTV IHXZG HDVIV ZOREV, is called the cyphertext. (Cypher is another word for a secret code.)

To decode the message, Bob just goes to row A in the table and reverses the operation from bottom to top to get back the message in plaintext. The

26 I'm not sure this is exactly the right protocol since the NSA tech support refused to take my calls – and without the instruction manual – I can't be sure.

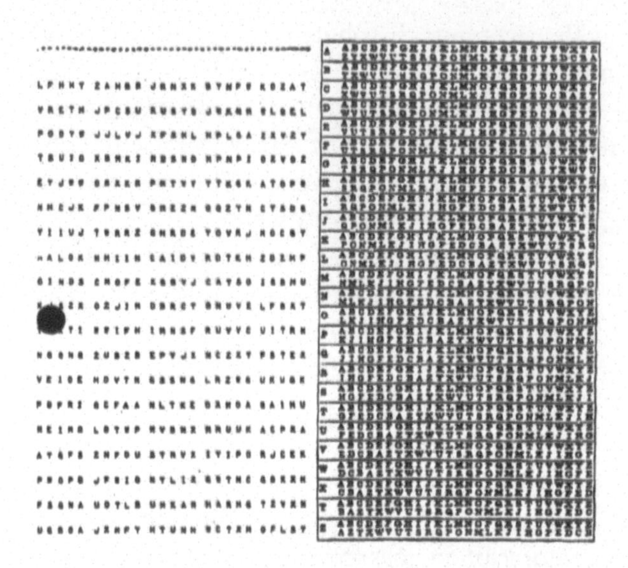

FIGURE 5.3 A form of a one-time pad used by the U.S. National Security Agency. The table on the right is an aid for converting a plain message into an encrypted message. In row A, you can read all the letters of the alphabet in order on the top and all the letters again in random order on the bottom.[27]

pad requires you to break up the message into five chunks of five letters each, as shown at the left of the figure.

Part of the protocol, set up in advance and agreed to by Alice and Bob, is that when a message does not go through, then they both move down to the next row B in the pad and try again – they must never use row A again! Alice now encrypts the message as follows, going top to bottom again.

S	C	H	R	O		D	I	N	G	E		R	S	C	A	T		S	W	E	R	E		A	L	I	V	E
G	W	R	H	X		V	R	Q	S	U		R	S	C	A	T		I	G	W	R	H		Y	N	Q	D	U

The plaintext SCHRO DINGE RSCAT SWERE ALIVE remains the same, but the encrypted cyphertext she sends to Bob is instead GWRHX VRQSU IGWRH YNQDU. As long as Bob also uses row B on the pad to decrypt, he gets the original plaintext message back by reversing the process.

27 This photo is the property of the U.S. Government and not subject to copyright.

In general, Alice and Bob agree to break up the plaintext into five-letter blocks such as SCHRO DINGE RSCAT. If I'm interpreting this NSA pad correctly (NSA tech support won't return my phone calls) – we do the encoding in five, five-letter blocks per row. We can see this structure on the left of Figure 5.3. To fill out five blocks, let's make the message now, SCHRODINGERS CATS ARE ALIVE. We break this up into five chunks of five letters per block (as indicated in the right of Figure 5.3) and then go to the lower part row A on the pad to encrypt, matching each letter in the message in the top to its crypto letter in the bottom.

Some of my more ambitious readers may have tried to take some of the cyphertext on the left of Figure 5.3 and then used the pad on the right to decipher it. I tried this myself but got gibberish. That probably means that the right part of the figure used a different pad, and somebody photocopied them together.

What's so bad about reusing the pad? Let's take an example. In many newspapers around the world, you can find a puzzle called the *Daily Cryptoquote*. Some famous quote is encrypted and published in code. The cryptoquote does not use a one-time pad but rather a so-called Caesar cypher – named after the emperor Julius himself.

The Caesar cypher is easy to break, and so are the newspaper Cryptoquotes. Let's break *this* one! The method is similar to that used in a popular U.S. television show called *Wheel of Fortune*. The contestants are allowed to guess a letter for the boxes in Figure 5.4. The smart players use the ETAOIN-SHRDLU method. The letters e-t-a-o-i-n and s-h-r-d-l-u are the most common letters in the English alphabet arranged in order, left to right, from most common to least. In the TV game, the first person to play almost always guesses an E. (Playing Wheel of Fortune is substantially easier than breaking the Cryptoquote, because in Wheel of Fortune, the game tells you when you got the letter right. In the Cryptoquote, you don't know that, and you have to keep trying over and over again.)

A brief inspection of Figure 5.4 indicates the most common letter is a B, which appears ten times. A good guess is that B corresponds to E, as shown in Figure 5.5.

The next most common letter in the Cryptoquote is T – with six total occurrences. We could try the most likely mapping, namely that T corresponds to T. We are suspicious of this – they never give away a letter in the Cryptoquote. We go next in the list and try that

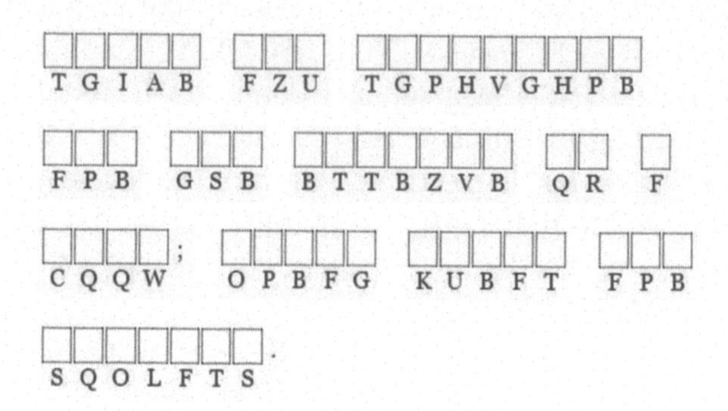

FIGURE 5.4 An example of a Cryptoquote. When decoded it becomes, STYLE AND STRUCTURE ARE THE ESSENCE OF A BOOK; GREAT IDEAS ARE HOGWASH, a quote by the Russian writer Vladimir Nabokov.

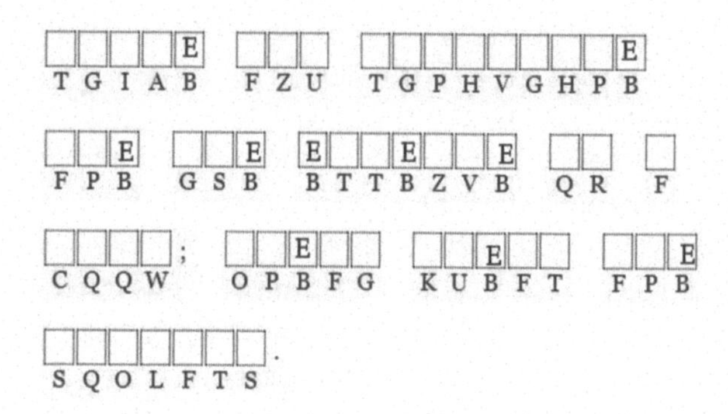

FIGURE 5.5 Our first guess is that B corresponds to the most common letter in the alphabet, E.

T corresponds to A. But immediately we are suspicious of this too, because if that is the correct correspondence, then EAAE_ _E must be an English word, which is highly unlikely. The same deal holds of we next try T corresponds to O. Then EOOE_ _E must be an English word. If T corresponds to I, then EIIE_ _E must be an English word. If T is N, then ENNE_ _E must be an English word. That is possible, but it seems unlikely. To check, I go to the dictionary and see if there is any seven-letter word that corresponds to ENNE_ _E. The answer is no. The dictionary goes from ENNEAD to ENNEAGRAM. That guess cannot be right. Next in line is that T corresponds to S. If true, then ESSE_ _E must be an English word. There is only one seven-letter word that fits the bill, and it is ESSENCE, and T corresponding to S is likely the correct guess. This analysis gives us two more clues, Z corresponds to N and V corresponds to C. We thus put T for S, Z for N, and V for C everywhere, as shown in Figure 5.6.

We also know that the most common three-letter word is either THE or AND, but the most common one-letter word is A. That observation provides us with the clue that FZU is AND, and F is A. We will fill these in too, and that provides us F is A and U is D, which we also fill in. That delivers us that FPB is _ _E, which is likely ARE. Thus, we can also guess F is A and P is R and fill that in. From that, we can speculate that _DEAS is IDEAS, and that presents us that K is I, and we fill that in.

Once we do that, we postulate that QR is likely OF and that tells us that Q is O and R is F, and we pencil those in too. Now we have found all the letters E, O, I, A, N, S, R, and D in ETOIAN SHRDLU. Instead of guessing H, L, and U, we go back to visit T, which is the second most common letter. The most common letters in the encryption, which we have not yet identified are the Gs, with five occurrences. Hence, we suppose that G is T. We surmise that T_E is the other most common three-letter word THE. That forks out that S is H, and we arrive the result in Figure 5.6.

We are almost there! I consulted the dictionary for a word that fits STR_C_ _RE, but there are too many to guess. But we can see that

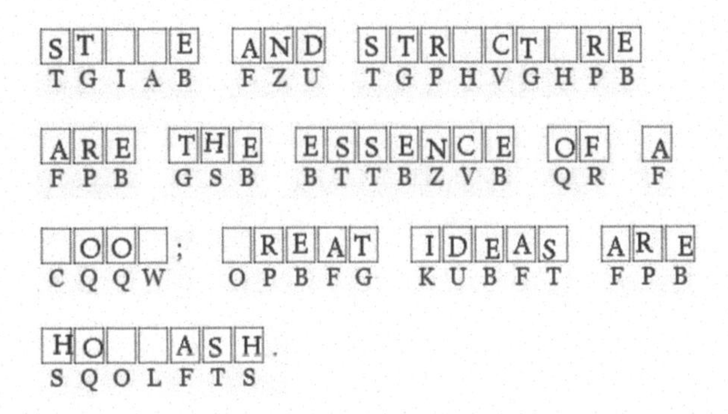

FIGURE 5.6 Our second guess is that T is S, which allows us to guess the word ESSENCE, giving us also that Z is N and V is C. We can postulate that QR is OF and F is A and that then FZU is AND, which gives us several more letters. We make more guesses using ETOAIN SHRDLU and that the most common three-letter words are THE and ARE.

REAT IDEAS must be GREAT IDEAS, and we get O is G, which allows us to immediately guess HOG _AS_ is HOGWASH, the only word in the dictionary that fits the bill. Now we get L is W, which does not help much since there is only one L. We can surmise STR_CT_RE is STRUCTURE, which gives H is U. At this point, we can reckon STY_ _E is STYLE, and _OO_ is BOOK – and we have solved the Cryptoquote![28]

The above procedure is either a very tedious process or lots of fun, depending on how you feel about Cryptoquotes. I find them easier to solve than a crossword puzzle, even though the crossword puzzle gives hints. But the Cryptoquote is much harder to solve than Wheel of Fortune, which tells you if you guess a letter right. The point is that it is solvable, given enough time, using knowledge of the English language. An NSA supercomputer can solve it in a second.

Let's now play the Cryptoquote using the one-time pad instead of the Caesar cypher. We can use an online random letter generator to do the encryption, as shown in Figure 5.7.[29]

The cyphertext below the plaintext is chosen from a one-time pad, similar to that in Figure 5.3. I have the pad – that's what I used to encrypt the plaintext – but you are the eavesdropper and don't have it. You could try the same approach as we did in the Cryptoquote, but you immediately run into big trouble. Recall that the one-time pad encryption

STYLE AND STRUCTURE
EQNYO JXY IAFWJLXJP

ARE THE ESSENCE OF A
BYQ COF GMMWKKD IO L

BOOK; GREAT IDEAS ARE
FTVH KCSHT ZXEAJ UEV

HOGWASH.
WNDKJNI

FIGURE 5.7 The same famous quote as before, but now encrypted with an unbreakable one-time pad.

28 This here was the first Cryptoquote I have ever bothered to solve in my life!

29 If you want to generate random strings of binary number yourself, go here www.random.org/strings/.

is secure only if we choose the letters at random – *we do not reuse the letters!* The letters in the Cryptoquote were chosen at random, but in that case, the pad was *reused!* This means that when we discovered that T in the cyphertext was equal to S in the plaintext, then *each* of the letters T corresponded to *each* of the letters S. Not so in the one-time pad. As we can see in Figure 5.7, the first time S appears in the plaintext it corresponds to E in the cyphertext, but the *second* time S appears in the plaintext it corresponds instead to the letter I in the cyphertext, and the *third* time S appears in the plaintext it corresponds rather to M in the cyphertext. There is no pattern! Using our knowledge of the English language, we were able to solve the Caesar-cypher-encrypted Cryptoquote, but that does not work with the one-time-pad-encrypted Cryptoquote. The fact that the letters from the pad are random and never reused is what makes the one-time pad encryption unbreakable. There is no pattern in a random string of nonrepeating letters. In the 1940s, Claude Shannon, the father of information theory, proved that the one-time pad encryption is unbreakable by any means – even by a quantum computer!

Why don't we just replace all of our quantum-computer-hackable public-key encryption around the world with unbreakable one-time-pad private-key encryption? Such a system would be immune to quantum computer attacks. There is one remaining problem. The third requirement for unbreakablity is that Alice and Bob must ensure that no eavesdropper has made a copy of the pad when the pads were in transit. This is – literally – the weakest link. In classical physics, one can always make a copy of classical data. There is no way to be sure that an eavesdropper did not covertly make a copy of the two pads while they were in transit from Alice to Bob. For this reason, people seldom use the classical one-time pad – because of the difficulty in transmitting the pads securely. In the U.S., the pads are used by the CIA, the diplomatic corps, and to secure the nuclear launch codes.[30]

30 In May of 2018, I helped run a quantum technology workshop at the U.S. Air Force Global Strike Command, which is in charge of launching (or hopefully not launching) our nuclear missiles. I chaired all the sessions and gave the first lecture of the conference – an introduction to quantum technology. Everybody had assigned seats, and I was seated in the front row at the left hand of a three-star general – the commander of the Air Force Global Strike Command. To his right sat his two scientific assistants, whom he called the "docs", because they both had PhDs. When I asked if I was also going to be called "doc", the general replied, "Hell no – you're the prof!" When I got to the part of my talk on one-time pads, I made the comment that they are used to secure the nuclear launch codes. The general froze up and turned to one of the docs and asked, "Is

As discussed in my previous book, a version of the one-time pad was used in the Enigma machines that secured data transmissions to and from U-boat submarines in World War II. The pads were placed in the U-boats at the port, and then they were assumed to be secure as the submarines went about their business of torpedoing allied ships. Nevertheless, the British mathematician and father of classical computing theory, Alan Turing, and his team at Bletchley Park, U.K., were able to crack the Enigma code.[31] When the allies landed at Normandy on D-Day, they knew precisely where every German submarine was, thanks to Turing and the gang.

How was the Enigma code defeated? The Enigma machines violated one rule, their pads repeated. They repeated only once every few billion messages, but they did repeat. The Enigma machine users also violated another rule when they transmitted a message that did not go through, they often retransmitted the message again *using the same row of the*

---

that classified?" I leaned over and calmed him down by saying, "Sir, I don't know. I do not have a security clearance. But I did read it on Wikipedia."

That was a relief – and apparently – that knowledge was not classified. Out of curiosity, I asked the general how they delivered the one-time pads from the command center to the nuclear missile silos. In his strong Arkansas accent, he replied, "Well, prof, we have a guy in a truck that drives around to the missile silos and delivers the pads on three-and-a-half-inch floppies." Everybody in the audience gasped, including me. I then retorted, "Seriously!? – Three-and-a-half-inch floppies!?" The general's reply was priceless, "They don't make the eight-inch anymore!" American physicist Paul Kwiat – an endearing bird-like little man who is an expert on quantum cryptography – yelled from the back of the room, "We can do better than that!" The general replied, "Well nobody can hack a guy in a truck." I followed up with, "Yes – but somebody could drug or bribe the guy in the truck and copy the pad without anybody else knowing – with quantum cryptography we can be sure that nobody can make a copy and you can't hack that either." Then somebody in the back of the room asked, "How much would it take to roll something like that out?" I paused, took a deep breath, and then said something that I had never said before in my life, "At least a billion a year to start." I momentarily felt like Doctor Evil in the Austin Powers films when he names an outrageously large (or small) sum for ransom money – and everybody laughs at him. This time nobody laughed. The room was quiet for a moment, and then a doc said, "That seems reasonable," and wrote the number down in his notebook. This entire engagement has to be the most surreal moment of my life – the only thing standing between us and global thermonuclear annihilation – *is some guy named Buck in a truck!*

As I'm writing this book, things are changing so rapidly, that instead of going back constantly to re-write stuff, I'm footnoting the footnotes. In October of 2019, the U.S. Government announced that they were moving away from the 3.5″ floppies to a "highly secure, solid-state, digital-storage solution." See, www.businessinsider.com/military-replaces-floppy-disks-used-to-control-nuclear-weapons-2019-10. For all I know, this could be a USB stick. I *hope* the move was prompted by me teasing them in public about the floppies.

31 All about Alan Turing https://en.wikipedia.org/wiki/Alan_Turing. There is a very nice film titled *The Imitation Game* (starring Bandicoot Cummerbund) that tells this whole story – all the way up to when Turing takes his own life by eating a cyanide-laced apple rather than being sent to prison or enduring chemical castration for his "crime" of being a homosexual. (The fact that he was a war hero was classified.) See https://en.wikipedia.org/wiki/The_Imitation_Game.

pad – instead of moving to the next row. They reused the pad. (This is what happens when you tell somebody to blindly follow the rules without telling them why they must follow them. If they don't understand the "why", they will tend to ignore the rules.) And finally, a real give away was that they started every message with "HEIL HITLER". Since the pads (technically the Enigma machine rotor settings) were changed only once a day at midnight, for 24 hours the codebreakers knew the cyphertext corresponding the plaintext letters H, E, I, L, T, and R. That is, every day the Germans gave them 6 out of 26 of the cyphertext letters for free! Finally, the allies managed to capture an Enigma machine from a U-Boat, which told them a great deal about the settings and how often the pad repeated. Even with all this information, they had to build one of the first digital computers, called the Bombe, to break the daily code settings in time to extract useful information within the 24-hour window.[32]

If we are very cautious (and don't make mistakes like the Germans and reuse the pad) and if the characters on the pad are darn-tooting random, and no adversary makes a copy of the pad, then the scheme is unbreakable by any means – even by a quantum computer. There are really two problems with getting this all correct classically: we cannot classically ensure that any sequence of numbers is really random and we cannot classically ensure that the pads are not copied by an eavesdropper in transit. Both these problems can be solved with quantum technology. The QRNG certifies that the random characters on the pad are real-deal random. All that remains is to use a second quantum technology that prevents the eavesdropper from copying the pad. That second quantum technology is technically called quantum key distribution, but most people just refer to it as quantum cryptography. In any quantum network, like in any classical network, we must be sure the network is secure against eavesdroppers. Since it is a quantum network, we have no need for the public-key encryption, and we will use a quantum private-key encryption instead. There are numerous QKD protocols, but I will

32 How I learned to stop worrying and love the bombe https://en.wikipedia.org/wiki/Bombe. We are not sure why it is called the bombe, which is the Polish word for bomb. The most likely reason is that the machine made a ticking noise like a bomb. Another interesting (and perhaps apocryphal) story about the danger of reusing the pad goes as follows. The two communist revolutionaries, Che Guevara and Fidel Castro, communicated via a one-time pad – that much is true. The apocryphal part is that, as supply lines were cut, Guevara began running out of pad and began to reuse it. It was then that the CIA was able to decode it, locate Guevara, and have him taken out and shot.

discuss the two that are most likely to be used in the quantum internet in the near term.

QUANTUM MONEY – WIESNER'S UNPUBLISHABLE PAPER

Before there was quantum cryptography, there was quantum money. The godfather of quantum information theory is a little-heard-of fellow by the name of Stephen Wiesner.[33] Two out of the three elements of the ingredients of quantum weirdness – uncertainty and unreality – Wiesner put forth in a paper he wrote in 1970. Unfortunately, the stuff in Wiesner's paper was so radical, and Wiesner was such a non-self-promoter, the article was not published until 1983.[34] The idea is to encode the unique serial number of a 100-dollar bill in quantum states embedded into the currency (Figure 5.8). Let us say that the binary serial number is 10101010, which is printed on the face of the c-note.[35] (In information theory, eight bits is a byte. This is a one-byte serial number.) The problem is that anybody with a good photocopying machine could duplicate this bill and have *two* hundred-dollar bills. It would take a long time for the bank to

FIGURE 5.8 Quantum money. The serial number on the bill is encoded quantumly into quantum spin or polarization states buried inside the bill. Any attempt to copy the quantum states destroys the serial number. The copied bill now has a different quantum state, which can be detected by the bank.

33 A short bio of Wiesner is here https://en.wikipedia.org/wiki/Stephen_Wiesner.

34 More about quantum money https://en.wikipedia.org/wiki/Quantum_money.

35 This image is license by Taylor & Francis publishers from Shutterstock https://image.shutterstock.com/z/stock-photo-one-hundred-dollars-bill-with-no-face-isolated-on-white-clipping-path-included-220417372.jpg.

figure out there were two bills with the same serial number in circulation. By then, the counterfeiter would be long gone.

Wiesner's fix was to also encode the same serial number quantumly inside the bill using photon polarizations. For each digit in the bill's serial number, the bank uses its QRNG to decide to encode the photon either in the H–V basis or the ±45° basis. To simplify the notation, the encoding in the H–V basis is $|\leftrightarrow\rangle = 0$ and $|\updownarrow\rangle = 1$ (in what I hope is an obvious notation), and the encoding in the ±45° basis is $|\swarrow\!\!\nearrow\rangle = 0$ and $|\nwarrow\!\!\searrow\rangle = 1$. There are eight digits in the serial number – so the bank needs eight random numbers: 00001011. These numbers are *not* the serial number, but rather the bank's private key used to encrypt and decrypt the serial number into and out of the buried photons. If the private-key digit is a 0, then the bank chooses the H–V basis to encode the corresponding serial-number digit, and if the private-key digit is a 1, then the bank uses the ±45° basis to encode the corresponding serial number digit. The whole scheme is illustrated as follows:

$$|\updownarrow\rangle_1^{H-V}|\leftrightarrow\rangle_0^{H-V}|\updownarrow\rangle_1^{H-V}|\leftrightarrow\rangle_0^{H-V}|\nwarrow\!\!\searrow\rangle_1^{\pm 45^\circ}|\leftrightarrow\rangle_0^{H-V}|\nwarrow\!\!\searrow\rangle_1^{\pm 45^\circ}|\swarrow\!\!\nearrow\rangle_0^{\pm 45^\circ}.$$

Here the notation is that the subscripts are the serial number, and the superscripts are the basis encoding of the random numbers. The serial number is public – it is printed on the front of the 100-dollar bill. But the basis encodings are private – only the bank knows them. After the bill is in circulation and then is returned to the bank, the bank can check if it is counterfeit or not by measuring each of the photons one-by-one in the same basis that the photons were encoded into in the first place. Such a measurement has no effect on the photon polarizations and the bank confirms that it got back what it sent out by reading off the bottom row of subscripts to see if they match the bank's numbers.

What happens when a cantankerous counterfeitrix named Contessa tries to make a copy of the bill? Contessa has no access to the bank's private key. If she wants to copy the bill, she will need to copy the photon polarization states exactly as they are. But here is where quantum unreality and uncertainty thwart her dastardly plan. Classically, to copy something, you must first make a measurement of it and reproduce it. If I have a classical string of bits, such as the serial number 10,101,010 printed on the bill, the copy machine just duplicates them. However, again to uncertainty and unreality, it is impossible to copy an unknown quantum state, such as the states of the photons in the bill.[36] Contessa's best strategy is to try to measure all the photons – either all in the $H–V$ basis or all in the ±45° basis. Let us suppose that she chooses the former. Recall the encoding in the $H–V$ basis is $|\leftrightarrow\rangle = 0$ and $|\updownarrow\rangle = 1$, and the encoding in the ±45° basis is $|\nearrow\rangle = 0$ and $|\searrow\rangle = 1$. These are the numbers in the superscripts of the string of quantum states. With this, choice she will get the first four digits and the sixth digit of the serial number correct, as they were also encoded in the $H–V$ basis. However, the fifth, seventh, and eighth digit will get screwed up. Conti will be measuring in the wrong basis and will randomly collapse the polarization states of these serial number photons to either ↔ or ↕ with a 50–50 probability. That is because the +45° state $|\nearrow\rangle$ is unreal in the $H–V$ basis and collapses into either ↔ or ↕. One possible outcome would be

$$|\mathrm{C}\rangle = |\updownarrow\rangle_1^{H-V} |\leftrightarrow\rangle_0^{H-V} |\updownarrow\rangle_1^{H-V} |\leftrightarrow\rangle_0^{H-V} |\leftrightarrow\rangle_1^{\pm 45^\circ} |\leftrightarrow\rangle_0^{H-V} |\updownarrow\rangle_1^{\pm 45^\circ} |\updownarrow\rangle_0^{\pm 45^\circ},$$

which we compare to the unsullied bill,

$$|\mathrm{O}\rangle = |\updownarrow\rangle_1^{H-V} |\leftrightarrow\rangle_0^{H-V} |\updownarrow\rangle_1^{H-V} |\leftrightarrow\rangle_0^{H-V} |\searrow\rangle_1^{\pm 45^\circ} |\leftrightarrow\rangle_0^{H-V} |\searrow\rangle_1^{\pm 45^\circ} |\nearrow\rangle_0^{\pm 45^\circ},$$

where O is the original, and C is the copy. The states that now differ are colored blue in the original and red in the copy. Contessa can take the bill she attempted to copy and make as many more as she wants – but none of them will have the original quantum state encoded by the bank!

If the copy is taken back to the bank for verification, the bank will measure in the private encoding sequence of $H–V$ or ±45° bases and will discover that (with a 50% probability) the quantum bits for the serial number now differ in bits five, six, and seven (with another 50% probability). That is, the serial number printed on the bill no longer matches the serial number encoded in the photons. The bill is declared counterfeit by the bank. This scheme has the additional feature that Contessa not only cannot copy the bill, she renders the original bill useless. Instead of getting 200-dollar bills for the price of one, *she instead gets no 100-dollar bills at all!*

36 The inability to copy a quantum state is called the no-cloning theorem https://en.wikipedia.org/wiki/No-cloning_theorem, which was first proved in 1982. Recall, however, that Wiesner's quantum money paper was written in 1970. Several years ago, I gave a talk in Israel and Wiesner came to it. I asked him how he could have invented quantum money without the no-cloning theorem, and he told me that the theorem followed logically from the Heisenberg uncertainty principle and therefore required no proof.

FIGURE 5.9 A photo of me (solid blue shirt) and Stephen Wiesner (straw hat) taken at the Hebrew University of Jerusalem in June 2015, when Wiesner attended a lecture I was giving. If you look closely, you can see we are all fighting over a 100 New Israeli Shekel currency note – decidedly classical money.[37]

QUANTUM CRYPTOGRAPHY – THE BENNETT AND BRASSARD 1984 PROTOCOL

Charles Bennett, an American scientist who works at IBM, was a colleague of Wiesner and was fascinated with the ideas in Wiesner's quantum money manuscript. (See Figure 5.10.) Two out of the three requirements for quantum technology were both there, years before anybody had defined

37 The other people in the photo are Hagai Eisenberg (black shirt) and David Rohrlich (green shirt). As a joke, I pulled the 100 Israeli shekel note out of my pocket and suggested we use it as a prop for the photo. In the end, Wiesner won the tug of war and pocketed my money! I suggested that he should acknowledge me for funding in his next paper. Stephen Wiesner could never seem to find a permanent scientific job in the U.S. and immigrated to Israel decades ago to become a day laborer. This photo was taken using my camera by Lior Cohen, who has graciously transferred to me the copyright. To my knowledge, it is the only recent public photo of Wiesner in existence. The only other known photo was taken in 1968 www.mpiwg-berlin.mpg.de/research/projects/origin-and-development-quantum-cryptography.

what quantum technology was. In 1984, Bennett and Canadian scientist Gilles Brassard expanded on the ideas of Wiesner to produce the first paper on quantum cryptography. Since it was published in 1984, and the authors were Bennett and Brassard, the paper is universally referred to as BB84. I have discussed this protocol in agonizing detail in my previous book, and I'll only briefly review the topic here.[38]

As we discussed above, the one-time pad is absolutely unbreakable – even by a quantum computer – provided three criteria are met. You must not reuse the pad, the characters on the pad must be honest-to-goodness random, and you must ensure that an eavesdropper can never surreptitiously copy the pad. The first two demands are easily met, but the last is a problem – in the classical world somebody can always make a copy of the pad. This is the primary reason the pads are not more universally used because of this distribution problem. Quantum mechanics solves this problem via unreality and uncertainty – it is impossible to make a copy of an unknown quantum state – due to the quantum no-cloning theorem.

Let's recall just what Alice and Bob are distributing – shared random numbers in binary code. This is the private key. In the BB84 protocol, Alice generates a long random sequence of zeros and ones using her QRNG. The goal is to get this list of random numbers to Bob while ensuring that any eavesdropper (Eve) does not make a copy in the process. The idea is to encode the zeros and ones into the polarization states of photons. The procedure is very similar to the quantum money idea. After generating her private key, Alice uses her QRNG to encode the zeros and ones in either the $H-V$ basis or the ±45° basis. She chooses the $H-V$ if the random number is zero and the ±45° if the random number is one. Recall that the encoding in the $H-V$ basis is $|\leftrightarrow\rangle = 0$ and $|\updownarrow\rangle = 1$, and the encoding in the ±45° basis is $|\nearrow\rangle = 0$ and $|\searrow\rangle = 1$. Let's suppose the first byte of her private key is the random numbers 11,101,011. Now she generates a second set of random numbers for the encoding 01111011 that corresponds to bases choices $H-V$, ±45°, ±45°, ±45°, ±45°, $H-V$, ±45°, and ±45°. Hence, the encoded private key becomes

$$|P\rangle_A = |\updownarrow\rangle_1^{H-V} |\searrow\rangle_1^{\pm 45^\circ} |\searrow\rangle_1^{\pm 45^\circ} |\nearrow\rangle_0^{\pm 45^\circ} |\searrow\rangle_1^{\pm 45^\circ} |\leftrightarrow\rangle_0^{H-V} |\nearrow\rangle_1^{\pm 45^\circ} |\nearrow\rangle_1^{\pm 45^\circ},$$

38 For more on the BB84 quantum cryptoprotocol, see https://en.wikipedia.org/wiki/BB84.

where P stands for Alice's copy of the private key. This sequence of eight photons is sent over a public channel to Bob. Bob has his own random-number generator and randomly chooses – independently from Alice's choices – which basis to measure in. Say his random-number generator produces 00111000. This corresponds to Bob setting his polarization analyzers to $H-V$, $H-V$, ±45°, ±45°, ±45°, $H-V$, $H-V$, and $H-V$. The critical point to note is that 50% of the time, Bob chooses a different basis to measure in than what Alice decided to send in. That means he mucks up about half of the private-key bits. For this example, Bob's measurement produces

$$|P\rangle_B = |\updownarrow\rangle_1^{H-V} |\leftrightarrow\rangle_0^{H-V} |\nwarrow\!\!\searrow\rangle_1^{\pm 45^\circ} |\swarrow\!\!\nearrow\rangle_0^{\pm 45^\circ} |\nwarrow\!\!\searrow\rangle_1^{\pm 45^\circ} |\leftrightarrow\rangle_0^{H-V} |\updownarrow\rangle_1^{H-V} |\updownarrow\rangle_1^{H-V}.$$

We can see that Bob measured three of the photons in the wrong basis (colored in red). Clearly, for the second photon from the left, Alice sent a one, but Bob got a zero. For the last two photons on the right, Alice sent both ones, and Bob got both ones, but that is a complete accident. He could have just as likely gotten both zeros. The collapse is random.

Alice and Bob now have a way to discard the bad bits of the private key. They announce over a public channel, photon by photon, what basis Alice used to send and what basis Bob use to receive. Critically, they do not announce which bit was sent or received – but just the basis. Immediately Alice and Bob see that – from left to right – the three bits two, seven, and eight are possibly wrong because the send and the receive bases were different. They throw these results away and obtain a smaller private key of, in this case, five shared random bits. (It will be 50% shared random bits on average.) Discarding the red and blue bits and retaining the rest, Alice and Bob now share the same random private key,

$$|P\rangle_{AB} = |\updownarrow\rangle_1^{H-V} \otimes |\nwarrow\!\!\searrow\rangle_0^{\pm 45^\circ} |\swarrow\!\!\nearrow\rangle_0^{\pm 45^\circ} |\nwarrow\!\!\searrow\rangle_1^{\pm 45^\circ} |\leftrightarrow\rangle_0^{H-V} \otimes \otimes,$$

where $\otimes$ denotes a deleted bit. Translating this back into bits, the shared private key is 11010. Alice and Bob use this key to transmit a message via the one-time pad method.

What prevents an eavesdropper from copying the key? As with the quantum money, Eve's best strategy is to intercept and resend the photons in one basis, say the $H-V$ basis. If she does this, she collapses about half of the photon polarization states to the wrong value,

$$|P\rangle_E = |\updownarrow\rangle_1^{H-V} |\leftrightarrow\rangle_0^{H-V} |\updownarrow\rangle_1^{H-V} |\updownarrow\rangle_1^{H-V} |\updownarrow\rangle_1^{H-V} |\leftrightarrow\rangle_0^{H-V} |\leftrightarrow\rangle_0^{H-V} |\leftrightarrow\rangle_0^{H-V},$$

which disagrees with what Alice sent on photons two, three, four, five, seven, and eight. Now Bob will make his measurement on the state Eve sends to him, and whenever his basis is in the ±45° basis he will randomly collapse the states Eve has perturbed to get $H-V$, $H-V$, ±45°, ±45°, ±45°, $H-V$, $H-V$, and $H-V$. Consider that Bob measures the third photon from the left in the ±45° basis and it collapses to $|\nearrow\!\!\!\!\swarrow\rangle_0^{\pm 45^\circ}$. This outcome implies that Alice transmitted a *one* for this photon, but Bob got a *zero*! That cannot happen unless somebody is tampering with the system! What Alice and Bob do next is take about half of their private-key bits and declare them to each other over an insecure open channel. (They keep the other half secure for the private key.) If the openly declared key bits agree, then nobody is tampering with the quantum channel. If they disagree, then they don't use any of the key at all, for they know there is an eavesdropper, and they send out a security guard to take her out.

The primary thing that secures this protocol is the no-cloning theorem. Eve does not know what state Alice is sending. The theorem prohibits her from copying an unknown quantum state. Again, unreality and uncertainty kick in – the photon polarization state is *unreal* until a measurement is made – and then it collapses at *random!* Thus BB84 takes care of the final problem of distributing the private key in a way that ensures that nobody can make a copy without revealing themselves.

The ideal BB84 protocol should be carried out with single photons. However, the single-photon source technology is still slow, expensive, and cumbersome. Many deterministic single-photon guns require the source to be refrigerated to near absolute zero. Instead, the community has turned to using faint laser pulses.[39] Laser technology is well advanced, and it is relatively easy and cheap to make attenuated laser pulses with very rapid repetition rates for high-speed quantum cryptography. The pulses are time-binned so that in each picosecond time slot, there is at most one pulse. The problem with faint laser pulses is that they do not contain exactly one photon – they sometimes contain two or more – and they often contain zero. (See Figure 4.5.) Let us suppose we have time-binned the

39 For more on single-photon sources from faint laser pulses, see https://en.wikipedia.org/wiki/Single-photon_source#Faint_laser.

laser pulses – we get one pulse per picosecond. That corresponds to a gigahertz transmission rate. If the pulses contain more than one photon, then Eve can put a non-50–50 beam splitter in the laser path and occasionally split off that extra photon, which gives her some information about the private key. To mitigate this beam splitter attack, Alice chooses laser pulses that are so weak that, most of the time, the pulse contains zero photons. Sometimes it includes one, and with a minuscule probability it includes two or more. Decreasing the odds of two or more increases the odds that the pulse contains zero photons, which reduces the key transmission rate. Even then, Eve still can be sifting off the very few extra photons in the pulses that contain two or more. It is impossible for Alice and Bob to detect Eve because photons are lost all the time in any communication system. Eve is hiding in the background photon loss.

The way to fix this is for Alice to randomly send – not a very faint private-key-bit pulse – but rather a very bright decoy pulse containing hundreds of photons. Since Eve does not know when these decoy pulses will come, she reveals herself to Alice and Bob because she removes many tens of photons from the decoy pulses – much more than can be accounted for in the natural loss of the communication system. In this way, Eve is revealed, and Alice and Bob then do not use the key and send a security guard to round up Eve. As an analogy, suppose that Alice and Bob are trying to communicate in a very dark cave with these very faint pulses. Eve could be hiding in the dark – *stealing* photons! But then Alice – totally at random – turns on a searchlight. Eve is then caught because she cannot predict when to jump out of the way to avoid the searchlight.

The BB84 protocol is now in use across much of China, and in particular, it is transmitting private key in quantum local-area networks in Shanghai, Hefei, Beijing, and a few other large cities. These networks are connected to each other by a fiber trunk-line, which runs along the high-speed train track between Shanghai and Beijing. There are secure nodes at each train station. This ground-based fiber system has been up and running since 2016 and currently secures financial and government data transmissions. The Chinese have also demonstrated the distribution of the private keys using the Chinese quantum satellite Mozi. In this case, the attenuated laser pulses are generated on the satellite and beamed down to ground telescopes. Mozi first establishes a key with one ground station, say in Austria (Alice) and then it creates a second key with another ground station – say in Beijing (Bob). This has actually been demonstrated over a distance of 4,000 km between these two ground stations.[40]

In order for Alice and Bob to share a key, they and the satellite (S) carry out the following ritual. The satellite first establishes a key with Alice,

$$SA = 00111010\ 00010011\ 01011111,$$

which we recall is a string of random zeros and ones. Now the satellite sets up another key with Bob,

$$SB = 01111010\ 10000011\ 10001110,$$

which is a different string of random zeros and ones. Here we use 24 bits or 3 bytes of key. Then the satellite adds the two keys together on its end, using the binary-addition rules,

$$\begin{aligned} 0 + 0 &= 0, \\ 0 + 1 &= 1, \\ 1 + 0 &= 1, \\ 1 + 1 &= 0. \end{aligned}$$

This is called binary arithmetic, which is like adding times on a clock. If we use a 24-hour clock, then 24:00 + 01:00 = 01:00 and not 25:00. In the same way 1 + 1 = 0 and not 2.

We make an addition table following these rules,

$$\begin{aligned} &00111010\ 00010011\ 01011111 \\ +\ &01111010\ 10000011\ 10001110 \\ =\ &01000000\ 10010000\ 11010001. \end{aligned}$$

We have that SUM = 01000000 10010000 11010001 is equal to the binary sum of the two private keys. Note that since the keys SA and SB are entirely random, their sum is also completely random. The satellite now broadcasts this sum over a public channel with radio waves. Anybody can make a copy of this broadcast. However, because it is completely

40 See https://phys.org/news/2018-01-real-world-intercontinental-quantum-enabled-micius.html.

random to an outside observer, only Alice and Bob can extract any information from this broadcast. When Alice gets the SUM from the satellite, she performs the following operation – she takes her private key SA (which she is keeping secret) and adds it to the SUM broadcasted from the satellite,

$$\begin{aligned} SA &= 00111010\ 00010011\ 0101111 \\ + SUM &= 01000000\ 10010000\ 11010001 \\ &= 01111010\ \ 10000011\ \ 10001110 \\ &= SB! \end{aligned}$$

That is, this operation reveals Bob's private key to Alice (and Alice alone). Now Alice knows Bob's private key that he established with the satellite. I will leave the next steps as an exercise to the reader, but in the same fashion you can show that SB + SUM = SA! That is, Bob takes the bits in the publicly broadcasted SUM, adds them to his private key SB (that he established with the satellite), and out pops Alice's private key SA. Alice and Bob now have two sets of the private keys to use to communicate via a one-time pad. In the actual experiment, Mozi first establishes the key SA with Austria, then waits until it is over Beijing, and then establishes the key SB.

It then performs the addition protocol, given above, and broadcasts the result to both Austria and Beijing. The private key was used to set up a secure video conference between the President of the Austrian Academy of Sciences and the President of the Chinese Academy of Sciences. While this was a heroic experiment, the next Chinese plan is to launch a fleet of nanosatellites that only carry out BB84 between ground stations. In this way, they intend to *secure their entire internet* with quantum cryptography! Within a few years, the entire country of China will be communicating internally with an unbreakable cypher *that cannot be cracked!* – even on a quantum computer! The country will go dark – and no matter how many supercomputers the NSA strings together – they will not be able to hack *that.* One final thing to note, the satellites in this protocol know everything. Hence, if I did not want the Chinese to tap my communications, I would certainly not use their BB84 satellite swarm for QKD. But with the Ekert protocol – coming up next – I certainly *would* use their satellites.

FIGURE 5.10 Photo of Charlie Bennett – the first B in BB84 – singing "Waltzing Matilda" in a karaoke bar in Tokyo, Japan, in September of 2010.[41] He is seated next Hiroshi Imai (left) from the University of Tokyo.

QUANTUM CRYPTOGRAPHY – THE EKERT 1991 PROTOCOL

In 1991, Artur Ekert[42] developed a new type of QKD protocol, which according to the naming convention, is E91.[43] (See Figure 5.11.) Ekert was unaware of the BB84 protocol at the time. The Ekert protocol exploits all three of the pillars of quantum technology, as it uses quantum entanglement and nonlocality in the generation of the private key. When Ekert showed his crypto scheme to John Bell, Bell exclaimed, "You mean this entanglement is actually *good* for something?" I covered this scheme in detail in my previous book – hint-hint! I'll go over it briefly again here.

41 I took this photo with my own camera and I, alone, hold the copyright. When I showed this photo to Chris Fuchs, a colleague of Bennett's, he exclaimed, "Charlie Bennett does not drink alcohol!" I replied, "Your evidence supports the hypothesis that Charlie Bennett does not drink alcohol with *you!*"

42 For a biography of Artur Ekert, see https://en.wikipedia.org/wiki/Artur_Ekert.

43 For more on the E91 quantum crypto protocol, see https://en.wikipedia.org/wiki/Quantum_key_distribution#E91_protocol:_Artur_Ekert_.281991.29.

In the future quantum internet, the network will allow users to share large amounts of entanglement. While BB84 with decoy states is an intermediate stopgap measure to protect our data from being hacked by the imminent arrival of quantum computers, it does not make use of entanglement, which will someday be a ubiquitous resource. Once that day arrives, we'll have something like the Department of Entanglement that controls and meters entanglement – just like the Department of Energy controls and meters electricity distribution on the power grid. In such a future, entanglement will be cheap or essentially free. The BB84 networks can do only one thing – allow users to share private keys. Once the quantum internet has entanglement spread all over it, you can do many things with that resource – private-key generation, teleportation, distributed quantum computing, distributed quantum sensing, and more! (These applications will be discussed in the next few chapters.) BB84 will someday end up in the communication history museums along with the Enigma machine, semaphore flags, and smoke signals. What will replace it will be the E91 – or some similar protocol – that exploits the magic of quantum entanglement.[44]

Almost everything I want to discuss here was covered already in chapter three. See, particularly, Figure 3.5 on the setup for the Bell-testing machine. Suppose that we adjust the source so that it always emits two entangled photons in the state $|\Phi^{+}\rangle_{AB} = |H\rangle_{A}|H\rangle_{B} + |V\rangle_{A}|V\rangle_{B}$. Because of the entanglement, if Alice gets an H, then Bob also gets an H. If Alice instead receives a V, then Bob also detects a V. Alice and Bob both assign $H = 0$ and $V = 1$. Because of quantum uncertainty and unreality, from shot to shot, Alice and Bob both randomly get either 0 and 0 or 1 and 1, with a 50–50 probability – and that outcome is unpredictable by any means. At the end of a three-byte run, they share, for example, the private key, $P_{AB} = 11111001\ 10111101\ 00110000$. Now the rest of the one-time pad protocol is run over any insecure classical communication channel as before. There are several innovative things to note in comparison to BB84. First, the key is not transmitted from

44 "Any sufficiently advanced technology is indistinguishable from magic." – Arthur C. Clark.

Any sufficiently advanced technology is indistinguishable from nonsense.

Jonathan P. Dowling

Alice to Bob but instead it is generated at the two detection sites. Unlike BB84, where Alice sends the key to Bob, here the key mysteriously appears at Alice and Bob's detectors – the key does not exist before then. Secondly, neither party needs a QRNG – the key generation itself is a QRNG. Thirdly, the first two properties imply that the source does not have a copy of the key as it would in BB84.

That last feature is a real winner for security. The Chinese quantum satellite Mozi has also recently demonstrated the generation of a shared private key between two ground stations in China using the E91 protocol and the entangled photon source on the satellite. I would happily use these entangled photons to generate a secret key, even if I worked at the CIA or the NSA, as there is a test to show that the satellite knows nothing – it is the Bell test. There is only one way that an eavesdropper between the satellite and the ground, or a spoofer in the spacecraft itself, might try to compromise the security of this protocol. It is important to note that, unlike in BB84, these spies cannot copy the key in transit because the key is not transmitted. The key does not exist until the photons are detected on the ground. The spoofer could replace the quantum entangled source with a classical pair of single-photon guns that she programs to emit pairs of photons randomly in H–H or V–V. At the ground, this would look the same as a result with the entangled photons. The eavesdropper could try something similar by intercepting the entangled photon pairs – from her lair in the basket of a hot-air balloon – and then send instead classically correlated photons in H–H or V–V with a 50–50 probability. We can catch both of these scallywags using the same approach. Randomly and without warning, Alice and Bob perform the Bell test on some of the arriving photons instead of using the photons to generate the key. If there is a spoofer or eavesdropper in the mix, then their actions are equivalent to classical hidden-variable theory gremlims! That means the Bell test will fail if the spies are present. When that happens, Alice and Bob will know the channel is compromised. The will then discard the key, and send a security guard in a dirigible to intercept the eavesdropper (or a cloaked killer satellite to take out the compromised source).

FIGURE 5.11 Selfie with Artur Ekert taken in September of 2019 at a banquet in Hefei, China, celebrating his winning of the 2019 Mozi Prize for promoting quantum technology. Ekert is on the left.[45]

POST-QUANTUM-CRYPTOGRAPHY – OFF-RAMP ON THE QUANTUM INTERNET

In 1977, Rivest, Shamir, and Adleman proposed the first public-key Encryption system, which now secures all of our financial data and much of our government data. Its security rests on one premise, "We don't think computers can speedily factor large numbers." While the jury is still out on whether classical computers can rapidly factor large numbers, we now know that quantum computers can. I was at a meeting on quantum technology on 5 July 2018 in Munich. There, a chief scientist from Intel informed us that their conservative estimate for a universal quantum computer was ten years. I asked, "Ten years or more?" She replied, "Ten

45 More on the Mozi prize http://miciusprize.org/. The name of the Chinese philosopher, for whom the prize is named, is Mozi in Chinese. The anglicized version is Micius. Since we no longer call Beijing "Peking" I see no reason to call the philosopher Micius. My anglicized last name is Dowling, but in the original Irish language it was Ó Dúnlaing.

years or less." Today the U.S. is ahead in the development of quantum computers. That lead will not last long. Due to that evanescent lead, the Chinese have decided to move to a quantum-cryptography-based system, BB84, which is immune to attacks by even a quantum computer. And they have committed billions of dollars to develop a quantum computer of their own.

In the meantime, the USA's response to this threat is *not* to invest in a provably secure quantum cryptosystem, but rather to move to a new public-key cryptosystem, post-quantum-cryptography. This is a new-fangled public-key system that they hope – but cannot prove! – is immune to an attack by a quantum computer. Post-quantum cryptography is the U.S. National Security Agency's proposed fix to the matter, and their protocols are being tested, quantified, and standardized by the U.S. National Institute of Science and Technology – ready soon to be rolled out. This approach is a fool's errand. This U.S. response to the threat posed by quantum computers is simplistic, nearsighted, and dangerous. Quantum-key distribution uses an unbreakable one-time pad. That pad scheme is used by the U.S. diplomatic corps, the CIA, and by the Air Force to secure the nuclear launch codes. Currently, the launch codes are distributed to the missile silos on 3.5″ floppies driven around by a guy in a truck. Surely, we can do better than that?[46]

And why are diplomatic communications, intelligence communications, and the nuclear launch codes secured with one-time pads? It is because the

46 For my screed on this matter, see http://quantumpundit.blogspot.com/2018/10/post-quantum-cryptography-is-off-ramp.html. For a more reasoned argument, see this article written by me and my brother Michael J. Dowling, https://quantum-computing.cioreview.com/cxoinsight/quantum-computing-dream-or-nightmare–nid-28757-cid-218.html. After reading these posts, several people tried to calm me down by telling me that I was exaggerating and that the National Security Agency and the National Institute of Standards and Technology (NIST) knew what they were doing. NIST is soliciting post-quantum cryptography public key algorithms from the general public. I asked a friend at NIST, who shall remain nameless for obvious reasons, how they were testing and screening these protocols. He said, "Our team first tries to prove that the algorithm can be hacked by a quantum computer. If we fail, we take the algorithm over to a guy in building XX and let him take a try – and if he can't hack it then we certify it as unhackable." If this sham continues, then in a few years all the data on the internet will be certified as secure by some guy in building XX at NIST. The worst-case scenario is that – instead of investing in real quantum cryptography – we instead invest billions of dollars on post-quantum cryptography.

Then along comes Peter Shor 2.0 who proves it is hackable on a quantum computer after all. What do we do then – invest in post-post quantum cryptography? After that it's posts all the way down. Meanwhile the Chinese are securing their entire internet with real quantum cryptography – provably unbreakable by any means. The U.S. thinks that real quantum cryptography is too expensive. How expensive will it be if we have to build it anyway, after our multi-billion-dollar investment in post-quantum cryptography fails, and the Chinese are hacking all of our data in the meantime?

users of these systems do not trust public-key encryption since it is not provably secure. Moving to a different unprovably secure public-key encryption system does not change this reality. Any policy that embraces post-quantum cryptography is a move away from the development of the quantum internet, for which no such public key is needed. The quantum internet is automatically self-secured by quantum cryptography – which is built in.

To quote myself, "The future of the quantum internet is in photons, and the short-circuiting of the development of optical quantum information processors in the U.S. means that the future quantum internet will have 'Made in China' stamped all over it."[47] The future of the quantum internet is *not* post-quantum cryptography. We have an utterly unbreakable QKD protocol – why the *hell* don't we use it!? Post-quantum cryptography is a band-aid on an arterial wound.

QUANTUM TELEPORTATION

Here we continue the theme that, one day, worldwide distributed entanglement will be a practically free and ubiquitous resource. This network of nodes sharing entanglement with other nodes will form the backbone of the quantum internet. As we have seen, in a sizeable entangled network, security comes built in. Parties connected to the network use the available entanglement to encrypt their data by deploying the Ekert 1991 protocol. The entanglement-enabled private-key distribution, combined with the resultant shared one-time pad, provides absolute security against any attack – even that by a quantum computer. It is all good and well to have an entirely secure network to transmit encrypted data. If that is all the network did, it would still be a pretty big deal. But shared entanglement is handy for so much more. In any network or computer architecture, one of the first problems that you run into is how to move data around on the network as quickly and efficiently as possible. In a classical network, there is only one solution – you make many copies of the data and then physically transmit the data bits from point to point over optical fiber, space relays, or even copper wire.

In the quantum world, numerous studies (including some of my own) have shown that the fastest and most efficient way to move quantum

47 Dowling, Jonathan P. *Schrödinger's Killer App: Race to Build the World's First Quantum Computer*, 173. Taylor & Francis, 2013. www.worldcat.org/oclc/829961807.

data is to teleport it. (I pause here until the giggling and Star Trek jokes subside.[48]) You heard me right - in the quantum world - we use teleportation! (See Figure 5.13.) Instead of moving the quantum bits (qubits) physically from node to node in the network, we can instead make a qubit disappear at one node and magically appear at another node. To do this, we must first set up a network where all the nodes share entanglement, but that is precisely what a quantum network is. Once free and easy entanglement saturates the network, we can begin to move the quantum data around without actually having to move it at all. Let's briefly overview how quantum teleportation works.[49] (As usual, more details are in my previous book.)

Consider the schematic diagram in Figure 5.12. The U-boat (U) prepares a single-photon polarization state $|\psi(\theta)\rangle_U = \cos(\theta)|H\rangle_U + \sin(\theta)|V\rangle_U$. The trig functions, which my older sister Ellen hates, are designed to let Ulf (the U-boat captain) pick a photon with arbitrary polarization. Since Ellen hates trig, as I'm sure many of you do, consider this notation as simply a list of possible states you can generate for different polarization angles θ. (I'll leave the proof to Ellen as a homework assignment.)

$$
\begin{aligned}
|\psi(0^\circ)\rangle_U &= |H\rangle_U, \\
|\psi(90^\circ)\rangle_U &= |V\rangle_U, \\
|\psi(45^\circ)\rangle_U &= |+\rangle_U, \\
|\psi(-45^\circ)\rangle_U &= |-\rangle_U.
\end{aligned}
$$

48 Whenever physicists talk about quantum teleportation, they say it is nothing like in Star Trek. They say this because they are not Trekkers. It is almost exactly like in Star Trek. When Captain Kirk is teleported to the surface, he himself is not teleported, but rather all the information needed to reconstruct him - his "pattern" - is teleported. It is the same as with quantum. The photon itself is not teleported but rather the photon's "state" is teleported, and the state contains all the information. The only difference is that Star Trek teleportation is classical. This conclusion can be seen by considering the episode in Star Trek Next Generation, where a transporter malfunction creates two copies of Commander Riker. That would be impossible with quantum teleportation as it would violate the no-cloning theorem. But the fact that the Star Trek transporter can create two Commander Rikers shows that it is not transporting Commander Riker himself but only the information needed to reconstruct him. Since classical information can be copied without error, it is perfectly sensible for the classical Star Trek transporter to create two Rikers. Normally the thing is programmed to create a new copy of Riker and destroy the old copy, so we don't have two of them running around with one growing a beard and joining a rebellion.

49 More on teleportation https://en.wikipedia.org/wiki/Quantum_teleportation.

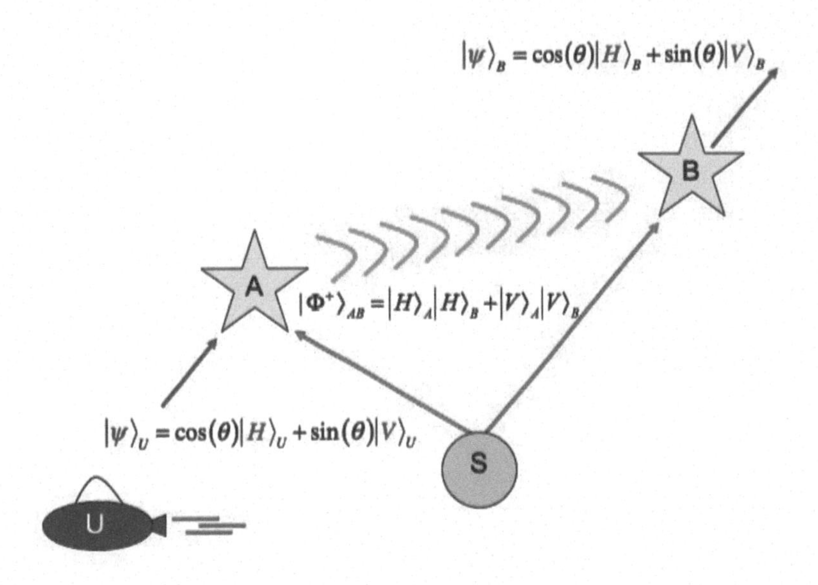

FIGURE 5.12 Alice on Alpha Centauri (A) teleports an unknown quantum state from the U-boat (U) to Bob (B) on Beta Pictoris.

Without even knowing trig, Ulf can dial up any of the states H, V, or ±45°, by just rotating the polarization angle. These are states we have used before but note that Ulf does not have to choose these four angles. He can choose any angle he wants and thereby choose an arbitrary state of the polarization, such as $|\psi(22.5°)\rangle_U = (0.92388\ldots)|H\rangle_U + (0.382683\ldots)|V\rangle_U$. This state is a quantum superposition of H and V polarization states with the state a lot more H than it is V. Another point to note is that the weight factors 0.92388... and 0.382683 ... are irrational numbers whose digits go on forever without repeating. To transmit this state classically from A to B, I need to use an infinite number of binary bits of information to describe the state and then communicate those untold numbers of bits from A to B (a tall order). But instead of doing this classically, we will use quantum teleportation.

For the sake of argument, let's say Ulf chooses this 22.5° polarized state to send. He knows what it is – but he does not tell Alice nor Bob. Their job is to transmit and receive the state without that knowledge. Classically, the best Alice could do would be to measure the state in some basis like H–V or ±45°, but since the state is not in either of these bases, her measurement would collapse the state into either H or V, thereby destroying all knowledge of the weight factors 0.92388... and 0.382683 ... Alice cannot measure the state, extract the weight factors, and then relay them to Bob. To do so would be to make a copy of the state. That would violate the no-cloning theorem. Even if it were not for the no-cloning injunction, it would take *an infinite number of classical bits* to send the weight factors 0.92388... and 0.382683 ... to Bob, because the numbers are infinitely long. Now, of course, she could relay the unknown state to Bob without measuring it, but it is a long way from Alpha Centauri to Beta Pictoris, and the state could become corrupted on the way. (More about this later when we discuss quantum repeaters.)

Another subtle point is that the weight factors 0.92388... and 0.382683 ... completely define the state. The state is not the photon but rather the quantum information carried by the photon. Instead of photons I could use nuclear spins and write,

$$|\psi(22.5°)\rangle_U = (0.92388\ldots)|\uparrow\rangle_U + (0.382683\ldots)|\downarrow\rangle_U,$$

where instead of photon polarizations, the up arrow is the nuclear spin pointing up and the down arrow is the spin pointing down. Even though the photon carries the state in the first case, and an atomic nucleus in the second case, the *state* is the same. When we talk about teleporting the state of a photon, we are not talking about teleporting the photon itself – but rather the *state* of the photon. I could readily teleport the unknown state of a photon into a state of a nucleus. The hardware platform – photons or nuclei – does not matter. It is the quantum information carried by that platform that matters – and that is what is teleported.

Alice now decides to teleport the unknown state to Bob. As in the Bell-test experiment, there is a source S that spews out entangled pairs, as shown in Figure 5.12. If there is no noise on the channels between the source and Alice and Bob, then one entangled pair will suffice. We select one of the Bell pairs, it does not matter which, and thus we take $|\Phi^+\rangle_{AB} = |H\rangle_A|H\rangle_B + |V\rangle_A|V\rangle_B$, which is the same we used in the E91 protocol above. As discussed in chapter three, Alice can now make a Bell measurement on two photons, her photon that is half of the entangled pair and the photon in the unknown state. A Bell measurement on two photons is the analog of a polarization measurement on one photon. For one photon, we pick one of the two bases H–V or ±45°. For two photons, we take one of the four Bell bases $|\Phi^{\pm}\rangle_{UA} = |H\rangle_U|H\rangle_A \pm |V\rangle_U|V\rangle_A$ or $|\Psi^{\pm}\rangle_{UA} = |H\rangle_U|V\rangle_A \pm |V\rangle_U|H\rangle_A$, where the notation indicates the two photons are the unknown one (U) and the one Alice got from the source (A). Again, in analogy, if we measure a single photon in the H–V basis, it collapses randomly into either the state $|H\rangle$ or the state $|V\rangle$. For the Bell measurement on two photons, the state collapses into one of the four Bell states written above. On top of that, the measurement tells Alice *which* of the four states it collapsed into. The measurement is destructive, and when completed, the state of both the U photon and the A photon is destroyed. That is, the information carried by the U photon is gone.

Where did it go!?

"When the going gets weird – the weird turn pro!"[50]

After Alice's measurement, Bob's photon on Beta Pictoris also collapses into one of four states with a 25% probability for each,

$$|\psi(22.5°)\rangle_U = (0.92388\ldots)|H\rangle_U + (0.382683\ldots)|V\rangle_U,$$

$$|\psi(22.5°)\rangle_U = (0.92388\ldots)|H\rangle_U - (0.382683\ldots)|V\rangle_U,$$

$$|\psi(22.5°)\rangle_U = (0.92388\ldots)|V\rangle_U + (0.382683\ldots)|H\rangle_U,$$

$$|\psi(22.5°)\rangle_U = (0.92388\ldots)|V\rangle_U - (0.382683\ldots)|H\rangle_U.$$

Notice that the first one is the unknown state – now sitting on Bob's photon! (The *photon* was not teleported the *state* was.) The other three are slight modifications of the unknown state. Which state Bob gets depends on which of the four measurement outcomes Alice got. Outcome one corresponds to the first state, outcome two to the second state, and so forth. Without this measurement data from Alice, Bob has no idea which of the four states he has. Alice now transmits only two classical bits of information: 00 if outcome one, 01 if outcome two, 10 if outcome three, and 11 if outcome four. When Bob gets these two bits from Alice, he performs the following operations on his photon, which can be implemented with beam splitters and phase shifters. If he gets outcome one he does nothing, if he gets outcome two he changes the minus sign into a plus sign, if he gets outcome three he

50 The quote may be found here www.brainyquote.com/quotes/hunter_s_thompson_100513.

replaces H with V and V with H, and finally, if he gets outcome four he replaces H with V and V with H and changes the minus sign into a plus.

At the end of these operations, no matter which of the four states he got, he now has the original unknown state $|\psi(22.5°)\rangle_U = (0.92388\ldots)|H\rangle_U + (0.382683\ldots)|V\rangle_U$ that currently resides on his photon and not Ulf's. The quantum information has been teleported, not the photon itself, that is, what we mean when we say the state has been teleported. Bob can now send the state into a quantum computer on his end to finish a quantum calculation that Ulf started on a quantum computer on his end. Or the state could be carrying a private-key bit as in BB84.

FIGURE 5.13 Photo of Richard Jozsa – one of the co-authors on the first quantum teleportation paper. I'm the guy to his right who looks like a pirate. You can see the back of Charlie Bennett's head to the left – he's the guy in the dark blue shirt. Photo was taken in Tokyo in August of 2010 at the Asian Conference on Quantum Information Science.[51]

51 The photo was taken in a Tokyo restaurant in 2011 by a Japanese geisha using my camera. She has signed over the rights to me in a note written entirely in Japanese on a sake-soaked napkin.

First important point: Teleportation *cannot* be used to send messages faster than the speed of light. Although the unknown state arrives at Bob's station instantaneously, it is garbled as it is one of the four possible states given above. Without the two bits of classical information from Alice, you can show that all that Bob gets on his end is random noise – and random noise contains no information. Bob must wait for Alice's two bits that travel over the classical channel at the speed of light – but no more. Teleportation comes itchingly close to blowing the theory of relativity out of the water, but in the end Einstein's theory survives by a hair.[52]

Second important point: The unknown state contains possibly an infinite number of bits of information in the weight factors. To send those weight factors classically would take an endless amount of time. Nevertheless, teleportation gets all those infinite quantum state bits from A to B – magically! – with only two additional bits of classical information transmitted from Alice to Bob. *How's that for data compression!?*

Third important point: There always must be a classical channel available to complete the teleportation protocol as well as the cryptography protocols. We assume that that channel is freely available and not the channel we send our quantum states over. For the space-to-earth E91 scheme, the classical channel consists of radio transmissions. For the BB84 trunk-line, between Beijing and Shanghai, the classical channel consists of the classical internet. However, it could be anything from semaphore flags to smoke signals. Initially, any quantum internet must have a classical internet backbone to complete the quantum protocols, but as we will see in the next chapter, this classical internet will be eventually subsumed by the quantum internet.

Experiments on quantum teleportation were first carried out in the late 1990s. The range has been extended over the years and the record is again held by the Chinese group of Jian-Wei Pan. His team teleported the state of a photon from a ground station in Tibet up to the Chinese quantum satellite Mozi – a distance of 1,400 km.[53] The previous record was held by the Austrian group of Anton Zeilinger who teleported the state of a photon about 150 km between two mountains in the Canary

52 My take on the weird interface between quantum theory and relativity can be found here http:// quantumpundit.blogspot.com/2013/09/on-curious-consistency-of-non.html.

53 Report on the Chinese teleportation experiment from Earth to Space www.space.com/37506-quantum-teleportation-record-shattered.html.

Islands.[54] Both of these experiments used photons flying through the air. For photons over an optical fiber, the record teleportation distance is only a bit over 100 km due to photons being absorbed.

ENTANGLEMENT SWAPPING

In the discussion of quantum cryptography and teleportation, we have assumed that entanglement is essentially free and distributed across a worldwide network. Any node in the network can share entanglement with any other node and thus carry out the E91 cryptography, teleportation, or distributed quantum computing. How does this entanglement get distributed all over the network in the first place? That is where entanglement swapping comes in.[55] Entanglement swapping is just teleportation that sends a *particular* unknown state – one half of an entangled state.

Consider the diagram in Figure 5.14. Alice and Bob would like to share entanglement, but they are too far apart for an entangled pair of photons from a single source to reach them. In any quantum communication channel, there is always loss, and particularly over optical fiber, the loss is quite severe. The longest you can get a single photon to go is about 100 km before the probability of is surviving is so low that the channel is useless. (More on this in the section below on quantum repeaters.)

Let us suppose Alice and Bob are 200 km apart – too far to share entanglement generated by a single source. Instead of one source S at the midpoint we place two entangled-photon sources S_1 and S_2 at the 50 and 150 km points, as shown in the figure. Midway between Alice and Bob, we place two new players, Charla (C) and Doug (D), who are colocated at the 100 km point and who have joint access to a Bell measurement machine. Charla and Doug do not have to be real people but just a box labeled Quisco that has beam splitters and detectors and other optical devices that can be run by remote control. The source S_1 generates an entangled pair of photons $|\Phi^+\rangle_{AC} = |H\rangle_A|H\rangle_C + |V\rangle_A|V\rangle_C$ and emits one photon to Alice and the other to Charla. Source S_2 spawns a second entangled pair $|\Phi^+\rangle_{DB} = |H\rangle_D|H\rangle_B + |V\rangle_D|V\rangle_B$ and

54 Article on the previous record for long-distance teleportation www.scientificamerican.com/article/entangled-photons-quantum-spookiness/.

55 Introduction to entanglement swapping https://en.wikipedia.org/wiki/Quantum_teleportation#Entanglement_swapping.

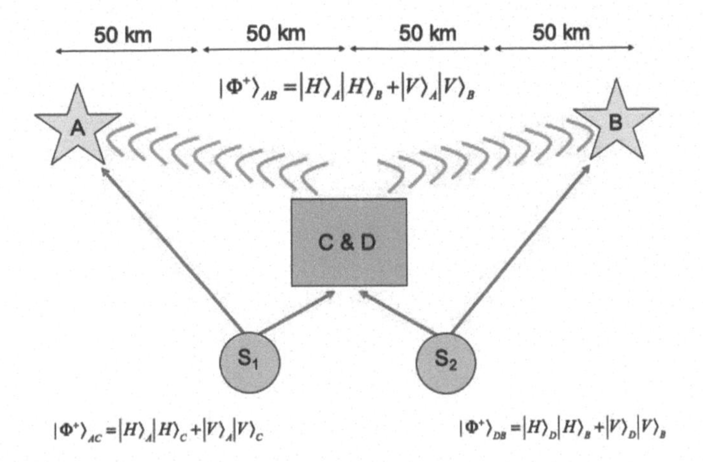

FIGURE 5.14 A diagram for entanglement swapping. The two sources S_1 and S_2 generate one entangled pair of photons each. The first source sends its two photons to Alice and Charla, and the second source sends its two to Doug and Bob. Charla and Doug make a Bell measurement on the two photons C and D and then transmit the outcome via radio waves to Alice and Bob using two bits of information with 00, 01, 10, and 11, which are the results of their measurement. Either Alice or Bob (or both) carry out one of the four polarization rotations indicated by the four bits. In the end, the C and D photons are destroyed, but the A and B photons are entangled. The protocol can be viewed as that the state of photon C was teleported to Bob or that the state of photon D was teleported to Alice.

transmits one photon to Doug and one to Bob. The way to think of this is to imagine that the C photon is in the "unknown" state that is to be teleported to Bob. Once *C* and *D* have their two photons, they do a Bell measurement on them – just as in teleportation. At the end of the measurement, the C and D photons are destroyed. But Charla and Doug know the outcome of the measurement – like in teleportation – it is one of four possibilities corresponding to the collapse into one of the four Bell states. Charla and Doug then relay the outcome to Bob, as before, encoding in the result into two classical bits, 00, 01, 10, 11. Alice and Bob now share the entangled state $|\Phi^{+}\rangle_{AB} = |H\rangle_{A}|H\rangle_{B} + |V\rangle_{A}|V\rangle_{B}$, as desired, over a distance of 200 km instead of the 100 km limit they would have had with just one source. Note that the *A* photon and the *B* photon did not come from the same source – nor were they ever anywhere near each other – and yet they are now entangled.

Entanglement swapping is part of the protocol that makes up a quantum repeater. Swapping alone cannot bridge a globe-spanning network. For now, let's stick to a ground-based fiber optical network like the one we primarily use now for the classical internet. Suppose we have a single fiber connecting Alice to Bob. For the typical commercial fiber, if Alice injects a single photon into one end, the chances the photon is lost hits 50% when the photon reaches the 7.6 km point. The probability that it survives the entire 200 km turns out to be exceedingly small – only about 0.00001%. That would be useless for a communication channel. By breaking the channel up into four parts, the farthest any photon has to go is only 50 km, as shown in Figure 5.14. The probability a photon survives 50 km is about 1%, which is much better. However, the swapping scheme only works *if all four photons survive.* The probability that all photons endure their independent 50 km trips turns out to be back at 0.00001%. It seems we have not gained anything by entanglement swapping. What we need is a quantum repeater that requires both entanglement swapping *and a quantum memory.*

QUANTUM MEMORIES

Quantum memories are ubiquitous requirements for nearly all quantum-information-processing tasks, particularly for quantum computing and communication.[56] In all of our discussions so far, we assume that, when an entangled photon pair arrives at Alice's and Bob's locations, they have some way of storing the state indefinitely. That storage is a quantum memory. But unlike a classical memory, which stores bits of information, the quantum memory must store qubits of information, maintaining quantum uncertainty, unreality, and entanglement. That is a much taller order. A poor-person's quantum memory is made by taking the incoming photon and put it into a fiber loop so that it just goes around and around until you need it. But that does not help since the fiber loop is as lossy as the transmission fiber. After a few round trips, the photon is lost. Another approach, one our group championed, is to put a type of quantum error corrector in the loop to refresh the state of the photonic qubit on each round trip. However, the technology for doing that is still years off.[57] Another approach is to transduce the qubit state

56 Lots about quantum memories https://en.wikipedia.org/wiki/Quantum_memory.

57 Gingrich, Robert M., Pieter Kok, Hwang Lee, Farrokh Vatan, and Jonathan P. Dowling. "An All Linear Optical Quantum Memory Based on Quantum Error Correction." Physical Review Letters

from the photon to another type of hardware platform - such as a trapped atoms - either confined in an atom trap or some solid-state material. One promising approach is to use an atomic qubit that sits inside a diamond where storage times on the orders of minutes have been seen.[58] (See Figure 5.15.) The biggest problem with these is that it is not a simple task to couple the photons in and out of the diamond, the diamond must be cryogenically cooled, and the wavelengths of photons that work with the diamond are quite a bit different than those that work well with fiber. The memory is long - but a lot of loss goes into getting the photons from the fiber into the diamond and back again. Once more, we point out, like in teleportation, the important thing is not the physical qubit that carries the quantum state, but rather the quantum state itself, regardless of whether it resides on an atom or a photon. For the discussion of quantum repeaters, we will assume that all the nodes in the network, such as *A*, *C*, *D*, and *B*, shown in Figure 5.14, possess a perfect quantum memory that can store the photons forever without loss or noise.

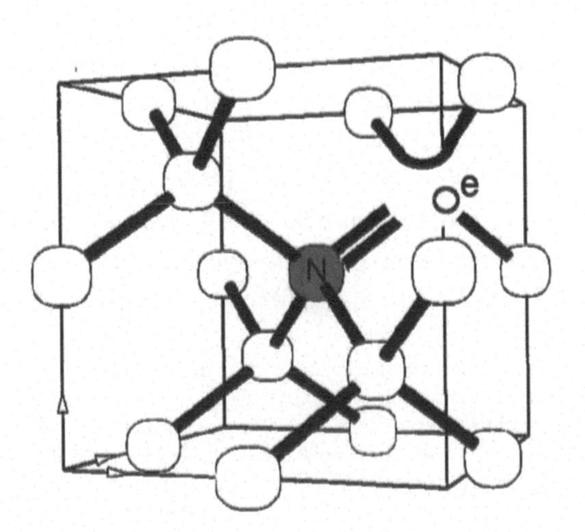

FIGURE 5.15 Stick model if a diamond quantum memory. One of the carbon atoms in the diamond is replaced with a nitrogen atom. The nitrogen atom distorts the surrounding atoms in such a way that it creates an atomic qubit that can reliably transfer in the quantum state from a photonic qubit and then back out.

Volume: 91 Issue: 21 Article Number: 217901 2003. https://journals.aps.org/prl/pdf/10.1103/PhysRevLett.91.217901 & https://arxiv.org/abs/quant-ph/0306098.

58 Diamonds are a quantum memory's best friend https://en.wikipedia.org/wiki/Nitrogen-vacancy_center & www.nature.com/articles/d41586-019-02766-3.

CLASSICAL REPEATERS

To understand what is so special about a quantum repeater, we'd best review what the heck a classical repeater is and how the classical internet was won. The internet was first something called the ARPANET[59] and was developed by what is now called the Defense Advanced Research Projects Agency (DARPA).[60] Figure 5.16 shows what the network looked like in 1974.

The ARPANET was developed to distribute information so that copies of important data existed at each node. If any few nodes were taken out in a nuclear attack, all the data would not be lost. The ARPANET was the idea of Paul Baran, and it goes back to 1959.[61] After some fits and starts, the network in Figure 5.16 was born. Important to note, for our discussion here, is that the data was initially transmitted over copper telephone lines. The transmission rates were very slow.[62] Transmission between continents was problematic due to loss in the copper wires, and often information or phone calls were routed via communication satellites.[63] The first undersea copper cable for use in telegraphy was put down across the Atlantic Ocean in the 1800s, but it only needed to transmit a few bits per second for Morse code. In modern communication schemes, we routinely transmit billions or even trillions of bits per second - copper wire can't do that. Instead, gradually, starting in the 1970s, the telephone companies began replacing the copper wires with high-speed optical fibers. A single strand of optical fiber can carry orders of magnitude - billions to trillions - more bits per second than a copper wire. Now, fiber-optic cables stretch across all of the continents and between continents on undersea cables. The copper

59 More on the ARPANET https://en.wikipedia.org/wiki/ARPANET.

60 More on DARPA https://en.wikipedia.org/wiki/DARPA.

61 More on the inventor of the internet https://en.wikipedia.org/wiki/Paul_Baran.

62 The first computer I ever used was when I was in high school in the 1970s - the machine was an RCA SPECTRA 70. The high school had a computer lab with a terminal that looked like an electric typewriter. I would use a dial telephone to connect to the mainframe computer in downtown Houston that the high school had a contract with. When the mainframe answered, making a squealing noise, I would take the headset of the phone and stick it into a phone-shaped indentation on the terminal. The bits and bytes of information were converted from electric signals to acoustic signals and then back into electric signals once more. The transmission rate of the system was 100 bits per second - slower than I could type! I would have to wait every couple of minutes for the thing to transmit the data out of a memory buffer before I could type again.

63 When I was a kid, I would call our grandmother in New York City, and about half of the time the call was fine, but on the other calls there was a weird delay in the conversation - we ended up talking as if we were using a walkie-talkie - pausing after we spoke to give the person on the other end time to hear what we said and to respond. Those delays occurred when the call was routed via satellite instead of copper wire. You could sometimes get around this by hanging up and calling back in the hopes the new call would be routed over the wires instead of the satellites.

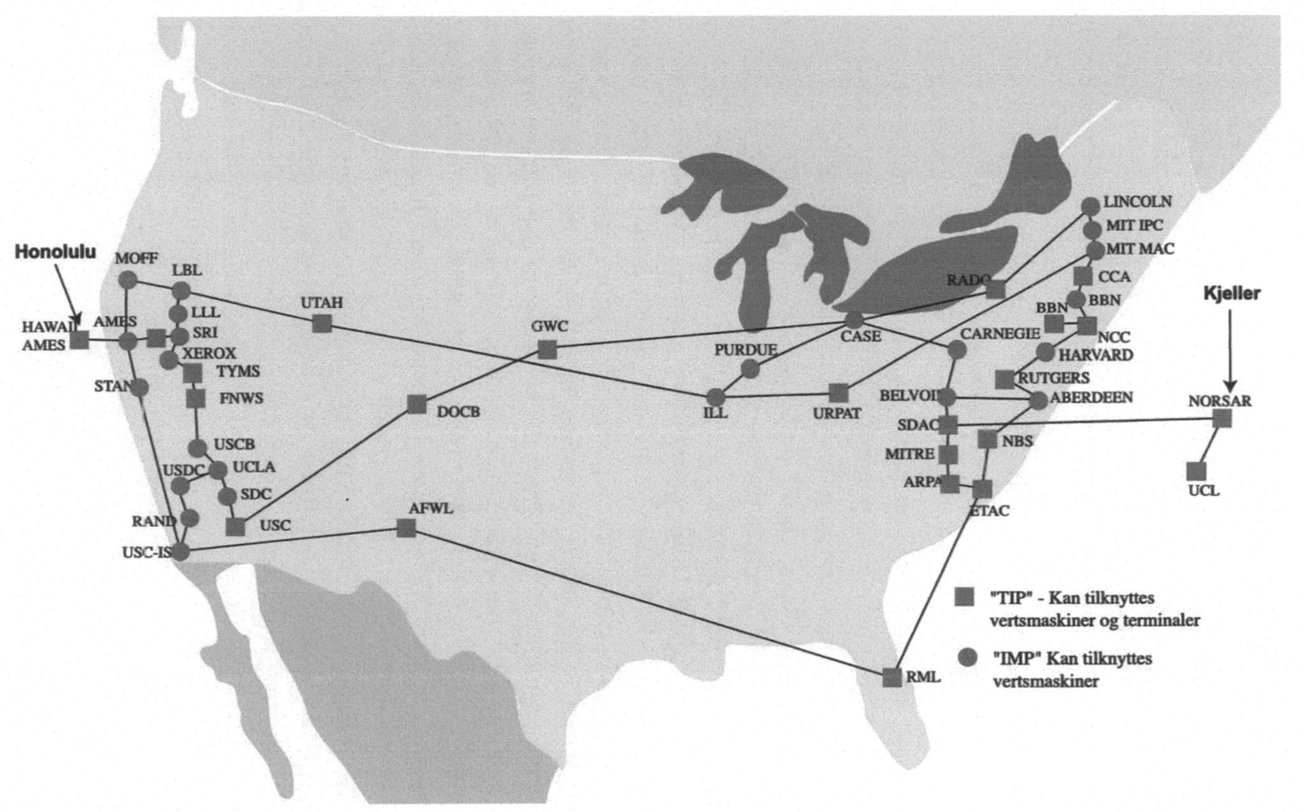

FIGURE 5.16 The ARPANET in 1974. It was the first internet connecting computers located at universities and government facilities.

wire has been mostly phased out.[64] Two breakthroughs made the fiber optic revolution possible – the development of very low-loss fibers and classical optical repeaters.

Let us suppose we want to transmit bits of information down an optical fiber at a rate of a billion bits a second. We attach the end of the fiber to a pulsed laser and then time-bin the laser pulses such that there is at most one pulse every billionth of a second, which is a nanosecond. We decide to encode the data as follows. If the laser emits a pulse in the nanosecond-wide window, we call that a bit 1. If it does not, we call that a bit 0. At the other end of the fiber is a photon detector that is synchronized with the laser. If it detects something in a nanosecond wide time bin, it registers a one, and if it detects nothing, it records a zero. In this fashion, the system can transmit a billion ones and zeros of information-carrying bits per second. These laser pulses are much brighter than discussed in BB84, and each pulse can contain a billion photons or more.

The problem is that even the best fiber has an exponential loss with distance. Using the numbers that we got in the section on entanglement swapping, if we send a pulse with a billion photons down 200 km of fiber, then only about 13 of those billion photons survive. Due to noise in the detector, while it can easily distinguish a billion photons from zero photons, it cannot distinguish 13 photons from zero photons. Hence, when it should record a one, it will most likely record a zero, and the information is lost. The fix is to break up the 200 km transmission line into stretches of 50 km. At each 50 km node, we put a classical repeater. The most widely used repeater is called an optical amplifier that is just a short stretch of fiber that is set up to act like a laser itself.[65] As the weakened information-carrying pulse enters this stretch of fiber, excited erbium atoms doped into the fiber emit photons in the same direction as the pulse. The atoms are excited by a separate laser that is located on the repeater node. There can easily be a billion erbium atoms in that fiber. You might imagine we could stick this at the end of the 200 km point, right before the detector, but the amplifier is very noisy – and it also amplifies the vacuum fluctuations.[66] Hence, it would amplify a pulse containing zero photons or a pulse containing 13 photons by the same amount, and we'd be unable to distinguish between the two.

64 Fiber-optic communication https://en.wikipedia.org/wiki/Fiber-optic_communication.

65 Doped fiber optical amplifiers https://en.wikipedia.org/wiki/Optical_amplifier#Doped_fiber_amplifiers.

66 In quantum theory, the vacuum is not empty, but contains a sea of fluctuating vacuum photons.

That's why we have to break up the transmission line into smaller 50 km links with repeaters at the intermediate nodes.

Suppose the initial laser pulse contains a billion photons. After a journey of only 50 km, it contains ten million photons. That number the amplifier can work with, and as the slightly attenuated pulse passes through the amplifier, it is brought back up to a full billion photons and then sent on its merry way. The same thing happens at the repeaters at the 100 and 150 km nodes. When the pulse arrives at the final detectors, it contains ten million photons, which the detector can easily distinguish from no photons at all.

The immediate question arises – why can't I use these classical repeaters to amplify the single photons or weak laser pulses that I need for quantum cryptography, teleportation, or entanglement swapping? There are two ways to view this at the quantum level: 1. The amplifier is noisy – that noise swamps the quantum signal destroying the qubit; 2. The amplifier is copying the incoming photons and using the copied photons to boost the signal. To amplify a quantum signal perfectly violates the no-non-noiseless amplification theorem, and to copy a quantum signal perfectly violates the no-cloning theorem. These two theorems are related, and if I could do either, then I could send signals faster than the speed of light, which is one way to prove the two theorems (by contradiction). Therefore, if I try to send quantum signals via photons over the classical internet, I destroy the quantum state being carried by those photons, which renders the communication channel useless. Copying or amplifying the quantum state destroys the unreality and uncertainty of the state, making it useless for a quantum technology that requires one or both of these features.

QUANTUM REPEATERS

Quantum repeaters include both the entanglement-swapping protocol and a perfect quantum memory at each repeater node. To see why the memory improves things, let us reconsider Figure 5.14. Without the memories, recall that the probability that all photons get through their independent trips of 50 km each turns out to be back at 0.00001% – too small to be useful. This number is gotten by assuming that all the photon arrivals are independent of each other. What this assumption implies is that the two sources must fire over and over again, to compensate for loss, and only when all four nodes, *A*, *B*, *C*, and *D*, get a photon is the swapping protocol implemented. That probability of getting all four is the 0.00001%. Most of the time at least one photon is lost and the swapping protocol must be restarted with the sources firing again.

The memory changes this assumption. With the memory – we can store two of the photons from the first source and then wait for the two photons from the second source to successfully arrive. Suppose Alice and Charla get a photon pair from the first source, and then they store each of their photons in their memories. If the memories are made of qubits in diamonds, then now the diamond qubits are entangled instead of the photons. Alice and Charla wait for Doug and Bob to successfully get an entangled pair to store into their own memories. All four of them know when all four of the photons have arrived, and then they transmit a classical signal confirming that all four photon states are in memory. Now they do the entanglement swapping between the diamond qubits instead of directly on the photons, and at the end of the repeater protocol, Alice and Bob share an entangled pair of diamond qubits, as desired.

The memory changes the odds of this all working. If we imagine the sources are spitting out a billion entangled pairs per second, then the four parties have a billion tries per second to get the thing right. What has changed with the memories is that once all the entangled qubits are verified to be in place, the swapping protocol now works 100% of the time and not just 0.00001% of the time. The tradeoff is in the source emission rate versus the fiber loss. The swapping works 100% of the time but Alice, Charla, Doug, and Bob get four photons only 0.00001% of the time, meaning their received rate is 0.00001% of a billion photons per second, which works out to Alice and Bob sharing an entangled pair of qubits about once per second. This rate does not sound good, but without the memory, they almost never share an entangled pair.

The goal is to improve the rate at which Alice and Bob share an entangled pair. The first improvement is to move the two sources onto the repeater node *C–D* in Figure 5.14. In that way, only two photons have to travel a long distance instead of four. The *A–C* photons and the *D–B* photons have to travel 100 km. Now Alice and Bob end up with entangled qubits at a rate of ten per second instead of one per second – an order of magnitude improvement.

The second upgrade is to break up the link into smaller links and add more repeaters say a total of three (instead of one) at the 50, 100, and 175 km points. There are now three entangled pairs (six photons) in play, but only four photons have to travel a distance of now only 50 km each. The probability that Alice and Bob share entanglement goes up again. In the limit that you put repeater stations very close to each other down the line, the probability that Alice and Bob share an entangled qubit pair grows to nearly 100%, and the swapping rate approaches that of the source – a billion pairs per second. Typical quantum network

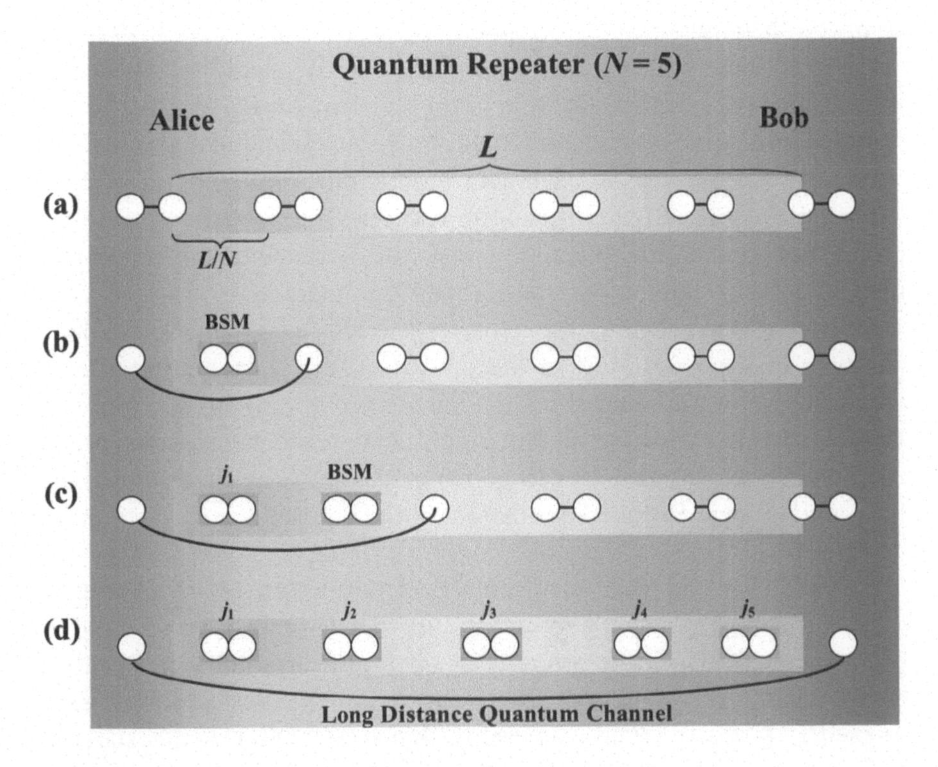

FIGURE 5.17 "A quantum repeater for entanglement distribution. (a) The transmission distance L is divided into N segments (here N = 5) with the intervening N–1 nodes as sources of entangled particles. (b) Midway on the first segment, a Bell-state measurement (BSM, blue box) swaps entanglement using one of the particles from node one and one of Alice's particles. The measurement outcome j_1 labels the newly formed quantum channel. (c) Repeated applications of the BSM extend the quantum channel further along the transmission line. (d) After N BSM's, Alice and Bob share a long-distance quantum channel determined by the string of outcomes j. Entanglement purification requires that steps (b) and (c) are repeated many times."[67]

67 "An Overview of Quantum Teleportation," by Travis S. Humble, https://web.ornl.gov/~humblets/publications/OverviewOfQTOctober2007.pdf. Travis Humble wrote this document as part of his duties as a U.S. Government employee, and as such, the work is not subject to U.S. copyright.

protocols have the repeaters stationed every 10 km, which gives about a million pairs per second – a very respectable rate.

In a quantum network, a repeater should be a box that contains a source of pure entangled photons produced at a high rate, a device for carrying out the Bell measurement (for the swapping), and a memory to hold the entanglement long enough for the entire string of repeaters to complete their mission. That mission allows Alice and Bob at the end stations to share entanglement at a high rate. A quantum repeater is a special-purpose optical quantum computer that runs a very particular protocol over and over again. Right now, superconducting quantum computers are leading the pack with atomic quantum computers close behind. However, all-optical quantum computers are beginning to come online and could rapidly leapfrog into the lead. Only the all-optical machines have a hope of interfacing with the internet.[68]

Critical for a network, the repeater should be small, cheap, and consume little power. Most of the optical quantum computing efforts are headed towards chip-sized devices where the swapping, photon source, and detectors are all on a chip that could be powered by a couple of small batteries. Not to be left out, I am on a team funded by the U.S. National Science Foundation to build exactly such a compact device.[69]

I should point out that quantum repeaters can also deploy entanglement distillation to purify photonic qubits that arrive at the end stations but are in bad shape. The property of quantum unreality is most often called quantum coherence. When I write the polarization state $|+\rangle = |H\rangle + |V\rangle$, the state is unreal in that it can either be a + state or a superposition of an H and a V state. In the parlance of quantum theory, this unreality is called quantum coherence. If the state interacts with a noisy environment, say polarization noise on the fiber, the photon might arrive at the end of the fiber but collapsed into H or V.

68 These statements have gone out of date within weeks of writing them. In November of 2019, Jian-Wei Pan's group at the University of Science and Technology of China demonstrated an all optical quantum computer competitive with the superconducting and atomic machines, www.scientificamerican.com/article/quantum-computer-made-from-photons-achieves-a-new-record/. In addition, also in November of 2019, the startup company PsiQuantum went public announcing millions of dollars of new investment into their approach to build an all-optical quantum computer, www.telegraph.co.uk/technology/2019/11/16/bristol-professors-secretive-quantum-computing-start-up-raises/. What is remarkable is that I'm typing this on Thanksgiving Day, 28 November 2019 – it is still November!

69 A summary of our team's effort to build a compact quantum repeater on a chip https://app.dimensions.ai/details/grant/grant.7733394.

The quantum coherence is gone – and we call this process decoherence. In optical fibers, this is a smaller effect compared to photon loss, but, nevertheless, it can be handled by a process called entanglement purification or distillation. If the state has not completely decohered, say it is still more than 50% in the original + state, you can take several these bad entangled states and distill them into fewer purer entangled states. In the limit that you have lots of bad states, you can get almost perfect states – but far fewer of them. The purification protocol requires that the bad states are stored in memories, and that additional quantum gates and classical communications are deployed. However, in most optical channel experiments, it is the loss and not the decoherence that is the most significant effect. The repeater protocol illustrated in Figure 5.17 shows both the original protocol to protect against loss along with the modified protocol to protect against decoherence.

QUANTUM TRANSPONDERS

There is a quantum repeater of the second kind that I will call a quantum transponder. It is also called a one-way quantum repeater or a quantum relay. It is a device that is much more like a classical repeater. Instead of relying on the entanglement-swapping protocol, the transponder takes a quantum state and applies a quantum error correction mechanism to the state periodically down the fiber. Our group designed the first quantum transponder protocol in 2003.[70] Since then, there have been published many extensions and improvements on our original idea in journal papers that don't cite me – but I'm not bitter.[71] Since we cannot copy or amplify the state, due to the no-cloning theorem, we take the original state and entangle it with additional helper photons.

In our 2003 transponder protocol, we take the two quantum states that we wish to transmit – each state encoded on a single photon – and then we entangle these two states with two additional photons call ancilla (helper) photons. It is a two-to-four encoding. The four photons can travel down the fiber together in different time bins in order to identify each of them

70 Gingrich, Robert M., Pieter Kok, Hwang Lee, Farrokh Vatan, and Jonathan P. Dowling. "An All Linear Optical Quantum Memory Based on Quantum Error Correction." Physical Review Letters Volume: 91 Issue: 21 Article Number: 217901 2003. https://journals.aps.org/prl/pdf/10.1103/PhysRevLett.91.217901 & https://arxiv.org/abs/quant-ph/0306098.

71 Borregaard, Johannes, Hannes Pichler, Tim Schöder, Mikhail D. Lukin, Peter Lodahl, and Anders S. Sørensen. "One-Way Quantum Repeater Based on Near-Deterministic Photon-Emitter Interfaces." https://arxiv.org/abs/1907.05101.

separately. Every 10 km or so sits a quantum transponder that checks to see if a photon is missing or not. If one is, the transponder performs a quantum error-correcting protocol by adding back in a new photon and then applying a sequence of quantum computing gates to regain the original four-photon state. It then sends this refreshed state down the fiber to the next transponder, and so on, until it arrives at the final node in the channel.

The transponder can be used to distribute entanglement as follows: Alice has a memory and an entangled photon source. After the source emits an entangled pair, she stores one photon from the pair in her memory, and she sends the other photon down the fiber. When the two photons arrive at the end station, Bob then stores these in his quantum memory. In the end, Alice and Bob share two entangled pairs in their memories.

The future quantum internet will allow all users (nodes) to share entanglement with any other user on the network. In this chapter, we have given the fundamental protocols that are necessary to make that happen. In the next chapter, we will discuss oodles of ways this entanglement distribution can be carried out and what practical applications would be right up the quantum internet's alley.

CHAPTER 6

Quantum Networks

[1]

Although this is not widely known, the American NSA-ARO-DARPA research programme in quantum computing was kindled, indirectly, by European speakers and communicated to the U.S. funding agencies by Jon Dowling who was, at the time, a physicist in the U.S. Army

1 Image from Shutterstock www.shutterstock.com/image-illustration/best-internet-concept-global-business-concepts-349563746.

Aviation and Missile Command …. Jon had attended the 1994 IQEC and ICAP meeting where British Telecom revealed their transmission of a quantum cryptographic key over 10 km of optical fiber, and Artur Ekert announced Shor's discovery of the factoring algorithm, respectively. Upon making these breakthroughs known to the Army Research Office, it was decided to have an ARO workshop on quantum cryptography and quantum computing in Tucson in the winter of 1995. Many of the key researchers in quantum computing were at this workshop, as well as Keith Miller and others from the National Security Agency (NSA). This workshop was the genesis of the ARO-NSA collaboration on the promotion of quantum computing research.[2]

THE QUANTUM INTERNET – WHY NOW?

In 2017, I wrote a book proposal for this book and sent it off to my publisher for review. The then sent it out to my peers for review. Several comments were along the line that we don't even have a quantum computer yet – it's too soon to be talking about a quantum internet. Well, that was 2017. Today several quantum computers have just hit quantum supremacy,[3] and everybody these days seems to be working on the quantum internet. A somewhat technical and out of date introduction to this topic is the book *Quantum Networking* by Rodney Van Meter.[4] This book is designed for quantum physicists who don't know much about how the internet works or for internet designers who don't know any quantum physics. In contrast, my current book here is designed for the technical lay reader, who knows nothing about either of these topics. Van Meter tells me that he hopes to update the book soon. There is also a nice Wikipedia page on *Quantum Networks* that first appeared in 2005, but that has since expanded rapidly.[5] This 2005 date precedes the first paper to be published on the quantum internet in 2008 by ℏ. Jeff Kimble.[6]

The critiques of my book proposal reviewers come to mind, why would we need a quantum internet before we had any quantum computers? Recall that the first classical computers appeared in 1945,

2 Brooks, Michael. *Quantum Computing and Communications*, 107. London: Springer, 1999 . www.worldcat.org/oclc/901463471.

3 For the definition of quantum supremacy, see https://en.wikipedia.org/wiki/Quantum_supremacy.

4 Van Meter, Rodney. *Quantum Networking*. Wiley, 2014. www.worldcat.org/oclc/904398228.

5 Quantum Network, https://en.wikipedia.org/wiki/Quantum_network.

6 Kimble, ℏ. J. "The Quantum Internet." *Nature* 453 (2008): 1023–1030 . www.nature.com/articles/nature07127, https://arxiv.org/abs/0806.4195.

but the ARPANET was not up and running until 20 years later. The ARPANET was built explicitly to hook classical computers together. However, right now, it looks like the first application for the quantum internet will be quantum cryptography. In 2016, the European Union launched its quantum flagship program with something called the quantum manifesto. In this document, they have a quantum technology timeline that puts the quantum internet before the development of a universal quantum computer.[7] Before the quantum internet they list quantum atomic clocks, quantum sensors, quantum intercity links, and quantum simulators – all to be connected with the quantum internet. The U.K. has a quantum initiative, launched in 2013, along the same lines with several millions of dollars invested in quantum technology hubs focused on similar topics.[8] China, for at least the past ten years, has been rapidly developing a quantum technology program of its own.[9]

The quantum internet will roll out in three stages. In quantum internet 1.0, partially up and running in China, the network does one thing – deploys quantum cryptography nationwide to secure its data. Their scheme uses BB84 with decoy states and has an operational fiber component linking cities from Shanghai to Beijing. This land-based network has been up and running since 2016 and currently carries financial and government data. The Chinese satellite has provided proofs-of-principle experiments illustrating that this key distribution can be deployed nationwide. By 2025, all of the government and financial communications in China will be secured with quantum cryptography – the country will go dark, and the U.S. will not be able to read anything. The Chinese have invested in this approach for two reasons. The first was that the Edward Snowden leaks showed that the U.S. had much greater classical eavesdropping capabilities than anybody expected. The second is that the U.S. will likely be the first to produce a quantum computer capable of hacking the classical internet – when that day comes, the Chinese will be ready for it. Quantum internet 2.0 will expand on this network and move from attenuated laser pulses to distributing entanglement across the network. It is unclear who will do this first – a tossup between the Chinese and Europe. (At this point, the

7 Quantum manifesto, https://ec.europa.eu/programmes/horizon2020/en/news/call-stakeholder-endorsement-quantum-manifesto.
8 UK Quantum Technologies Programme, http://uknqt.epsrc.ac.uk/.
9 Kania, Elsa B., and John Costello. *Quantum Hegemony*. Center for a New American Security, Washington, D.C., 2018. www.cnas.org/publications/reports/quantum-hegemony.

U.S. is playing catchup.) There will still be the need for a classical internet because, as we have seen, most of the interesting quantum protocols require classical channels to complete them. However, in 3.0, the classical internet will disappear entirely, since classical information can be transmitted over quantum channels at a faster rate than classical ones. In 100 years, we'll have an internet, and it will be all quantum.

WHAT'S UP WITH QUANTUM COMPUTERS THESE DAYS?

I am a founder of the U.S. program in quantum computing and quantum cryptography, and so I have a unique perspective on the U.S. program in these areas and also in quantum sensing. From 1995 until about 2015, nearly all of the funding for quantum computing development came from the U.S. government. While the names of the funding agencies changed over time, the vast majority of it came from the Department of Defense, and the vast majority of that funding came from the Intelligence Advanced Research Projects Activity (IARPA). The U.S. had put most of its eggs in a single basket – which allowed a lone hen to toss the basket off the roof. Around 2010, IARPA made a series of horribly informed executive decisions that crippled the U.S. program in quantum technology for years.

The first bad decision was – with about 30 days advanced notice – to stop all funding to U.S. government labs working on quantum computing and quantum cryptography. These projects included at least three programs at NIST and one at Los Alamos National Labs. The NIST projects were in atom-based quantum computing. Two Nobel Prize laureates ran the atom-based projects – William Phillips and David Wineland – and they each had won the Nobel Prize for work they had done at NIST. The head of the superconducting laboratory at NIST was John Martinis. He left NIST in 2004 after a previous round of U.S. government budget cuts and went on to become a professor at the University of California in Santa Barbara. At Los Alamos, two researchers, Jane "Beth" Nordholt and Richard Hughes, were working with NASA on a quantum communications satellite. Millions had already been spent on the project when, in 2009, IARPA killed it off entirely – a mere nine months before launch. Had IARPA not done that, the U.S. would now be ten years ahead of the Chinese satellite program, but instead the U.S. is ten years behind. The rationale behind this disaster was that IARPA management decided that the Department of Defense should not be funding programs run by other branches of the government – in this case the Department of Energy (Los Alamos) and the Department of Commerce (NIST).[10]

The second bad decision IARPA made was – not satisfied with the havoc they wreaked upon the government labs – to additionally kill off quantum computing programs not at government labs but at universities. I felt this sword first hand. In 2004, Gerard Milburn and I helped to put together a team to propose to an IARPA Quantum Computation Concept Maturation Program, to develop an all-optical quantum computer. The team had collaborators from around the world – from Australia to Austria. We won the grant and had four years of funding from 2005 through 2009 – a period in which we had quite promising results. At the end of the four years, we were strongly encouraged to submit a renewal proposal for an additional four years. We spent many person-hours on this proposal only to find, with little explanation, IARPA was not going to continue funding the work after all. (We wish they had told us that *before* we wrote the proposal.) Once again, entire programs in quantum computing were destroyed overnight. The word on the street was that IARPA had decided that optical devices were not a viable path toward a universal quantum computer.[11] After doing all that damage, IARPA then decided to instead fund a program of developing quantum compilers and programming languages that were designed to run on the quantum computers that IARPA was no longer paying to build.

With the above history in mind, the years between 2010 and 2015 were the dark ages in quantum technology in the U.S. The government funding practically dried up completely, and the U.S. program was kept on life support with the fumes of small grants from the National Science Foundation and smaller defense agencies, such as the Army Research Office. Some of the researchers survived on the chaff of former funding, but others left the government labs and academia to pursue quantum computing, in particular, in industry. When God closes a door, he opens a window – but he does not expect you to defenestrate your entire basket of eggs out of it. There is an old Russian saying, "Those who sit on the highest steps are the first to see the tsunami coming." John Martinis saw it in 2004 and left for the University of California.

---

10 Nordholt and Hughes – their program in ruins – took early retirement.

11 The word *off* the street was that the director of IARPA did not like something one of the principal investigators said to her at a program review and decided to cancel the entire program out of spite and revenge. After single-handedly destroying the quantum computing and cryptography, this particular IARPA program manager was promoted!

Then in 2014, Martinis took on a second position as a research scientist at Google, where he was given practically unlimited resources to build his dream – a scalable superconducting quantum computer. In his contract with Google, it explicitly states that he never will take a dime in funding from the U.S. Government ever again. The Google move paid off spectacularly when, in October of 2019, the Martinis team announced they had hit the Holy Grail of quantum supremacy – that is, they had built a quantum computer that could out-calculate the most powerful classical computers on Earth.[12] In another direction, Jeremy O'Brien and Terrance Rudolph, who were part of the IARPA defunded optical quantum computing program, left academia to start up a company called PsiQuantum to pursue the IARPA-abandoned dream of building a universal quantum optical computer.[13] There is also a trapped atom quantum computing company, spun off by the spurned program at NIST, called IonQ.[14] There are additional superconducting hardware projects at IBM and Rigetti Computing. Now we have worldwide tens of start-up companies either developing quantum computers or the software that will run on them.[15] When IARPA dropped the ball, industry filled in.

The U.S. is currently ahead of any other country in the area of quantum computing – not *because* of government funding, but in spite of the *lack* of it. However, other countries are catching up. Japan just announced a new multi-million-dollar investment in quantum computing and the Chinese – now that they have their quantum cryptography network almost complete – have been dumping money into quantum computing. They have a special-purpose optical quantum computer that hit quantum supremacy just a few days after Google did, and they are investing heavily in the superconducting approach.

---

12 In an amusing aside, when the announcement appeared, Ivanka Trump issued this Tweet,

> It's official! ✷ The U.S. has achieved quantum supremacy! In a collaboration between the Trump Admin, @Google, and UC Santa Barbara, quantum computer Sycamore has completed a calculation in 3 min 20 sec that would take about 10,000 years for a classical comp.

The irony of this announcement is that it is false – fake news! – Google did not take a dime from the Trump administration. See, https://twitter.com/ivankatrump/status/1186987509609385988.

13 "British Quantum Computing Experts Leave for Silicon Valley." *The Financial Times of London,* 22 June 2019. www.ft.com/content/afc27836-9383-11e9-aea1-2b1d33ac3271.

14 IonQ, https://en.wikipedia.org/wiki/IonQ.

15 A list of private or startup companies working on quantum computing https://quantumcomputingreport.com/players/privatestartup/.

The nightmare scenario for the U.S. is that the Chinese develop an internet-hacking quantum computer before the U.S. does but doesn't tell anybody. Their data will be secured by quantum cryptography, but the U.S. data is currently utterly vulnerable to a quantum-computing attack. Even if the Chinese quantum-hacking machine is ten years off, they could now be vacuuming up all of the U.S. data being transmitted over the internet and storing it on ginormous data farms. Then they will just wait and crack it in ten years when their own quantum computer is ready. There are financial, government, and medical data that you'd like to keep secret for decades – or if you are a Swiss bank – for centuries.

QUANTUM SUPREMACY

John Preskill coined the term quantum supremacy.[16] It means the point in time where we have a quantum computer that outperforms the most powerful classical computer on Earth. As of today, 30 November 2019, three groups have claimed to achieve this, Google, IonQ, and the photonic quantum computing group at the University of Science and Technology of China. There is some debate around these claims, most famously, IBM – Google's top competitor – retorted that Google had not hit quantum supremacy just yet.[17] This back and forth is but a quibble for the history books – 100 years from now those books will state that in 2015, we had not yet hit quantum supremacy yet, but by 2020, we surely had.[18]

What the *heck* happened?

In my 2013 book, I noted that many people thought quantum computers were decades away or would never be made. Suddenly here we are – only seven years later! – with a quantum computer that can outperform any classical machine. Quantum supremacy is important, not only as a technical milestone but as a philosophical one. Many computer scientists insist to this day that quantum computers are not any more powerful than classical computers. This leads us to a logical conundrum called Aaronson's trilemma.[19]

16 Preskill, John. "Why I Called It 'Quantum Supremacy'." *Quanta Magazine*, 2 October 2019. www.quantamagazine.org/john-preskill-explains-quantum-supremacy-20191002/.

17 Pednault, Edwin, John Gunnels, Dmitri Maslov, and Jay Gambetta. "On Quantum Supremacy." IBM Research Blog, 21 October 2019. www.ibm.com/blogs/research/2019/10/on-quantum-supremacy/.

18 "Quantum supremacy: the gloves are off!" by Scott Aaronson in his blog Shtetl-Optimized (October 2019), www.scottaaronson.com/blog/?p=4372.

19 "Schrödinger's cat is hunting masked chickens," by Scott Aaronson in his blog Shtetl-Optimized (12 June 2006), www.scottaaronson.com/blog/?p=92.

To quote Aaronson,

> I became interested in quantum computing because of a simple trilemma: either (i) the Extended Church-Turing Thesis is false, (ii) quantum mechanics is false, or (iii) factoring is in classical polynomial time. As I put it in my dissertation, all three possibilities seem like wild, crackpot speculations, but at least one of them is true! The question of which will remain until it's answered.

Let me translate and parse this a bit. The extended Church-Turing Thesis postulates that quantum computers are no more powerful than classical computers – a surprising number of computer scientists believe this – likely because they don't understand or want to understand quantum theory. The notion that factoring could be in classical polynomial time is the plot from the film Sneakers – there could be a classical factoring algorithm running on a classical computer that could hack the internet. There is no proof that there is *not* one, and computer scientists feel in their bones that this is exceedingly unlikely. (Let's secure the internet with a protocol which, if hacked, would make the computer scientists feel bad.)

The remaining possibility is that quantum mechanics is wrong – a conjecture sure to raise the hackles on any self-respecting quantum physicist. As a non-self-respecting quantum physicist, let me stake out our claim. Quantum theory has been tested in the crucible of experiment and no deviation between quantum theory and reality has ever been found – that in spite that the theory has been tested to many, many, many significant digits and over vast distances. If quantum theory is wrong, our entire vast enterprise of technologies that rely on quantum physics would be in peril. I do not merely *believe* that quantum theory is true; I have vast amounts of *evidence* to support that claim. Science is not philosophy or religion – I do not care what you *believe*, I only care what you can *prove*. Given the trilemma, if quantum theory is true, then either quantum computers are more powerful than classical computers *or* there exists a classical fast-factoring algorithm (or both). Quantum supremacy implies the former – I don't give a damn about the latter.

On the other hand, for a classical-computer scientist, who believes that quantum computers are no more powerful than classical computers, and who *also* believes there is not a fast, classical, internet hacking machine, the only option left for them is to conclude that quantum theory is *wrong*. This conclusion would be a delightful outcome for many of them – they never liked quantum theory in the first place and had no interest in learning it. I know classical-computer scientists who have spent a great deal of time arguing that quantum computers cannot be more powerful than classical ones.[20] Quantum supremacy means that such doubting Thomases are wrong.

A century ago, Max Planck, the founder of quantum theory, was asked, "How is the adoption of the new quantum theory coming along?" Planck replied, "Slow but sure – one death at a time." What Planck meant by this was that the old guard never accepted quantum theory – they eventually died – making way for a new generation of quantum physicists trained in the field. I can say the same now to the question, "How is the adoption of quantum computing coming along?" My answer is, "Slow but sure – one death at a time." The old codgers – who mortally cling to the Extended Church Turing Thesis – will soon all be dead. They will be replaced by young codgers, such as my high school students, who grew up with quantum computing and don't give a damn about the Extended Church Turing Thesis.

DOUBLE-BUBBLE – TOIL AND TROUBLE!

We hit quantum supremacy in the fall of 2019. How is that possible? In my 2013 book, I predicted that this milestone was decades away. In March of 2019, I attended a conference called *The Future of Quantum Computing, Quantum Cryptography, and Quantum Sensors*, held in Boston. During a panel discussion on the threat of quantum computers to such things as blockchain, a panelist said, "I don't see why we have to worry about a quantum computer that won't be ready for 100 years when we'll all be dead!" Here I sit, nine months later, writing about quantum supremacy. I can assure you that this panelist is still very much alive – only their business plan for securing blockchain against a quantum attack is dead.[21] How did we hit quantum supremacy so quickly? The answer is in quantum Moore's law.

As discussed in detail in chapter six of my previous book, Moore's law is the observation that the number of transistors on a computer chip has

20 Kalai, Gil. "The Argument against Quantum Computers." *Quanta Magazine*, 7 February 2016, www.quantamagazine.org/gil-kalais-argument-against-quantum-computers-20180207/.

21 Blockchain is a widely used cryptographic protocol that relies on quantum-crackable public-key encryption. See https://en.wikipedia.org/wiki/Blockchain.

been doubling every 18 months for the past 30 years.[22] Roughly, each transistor is equal to a classical bit of information. Hence, Moore's law states that the number of bits on conventional computers has been increasing at an exponential rate for 30 years. It is hard for humans to grasp how fast an exponential rate is. In the 1970s, I worked on a CDC6600 computer, one of the most powerful machines in the world at that time. This computer filled a large room, weighed over five metric tons, and cost two million dollars. Today I have a mobile phone in my pocket that costs $600 and weighs only 148 g. My phone has 64,000 times the memory and 250 times the processing speed of the old CDC machine. How is that possible? Moore's law is how.

In 2013, there were several front-runner quantum-computing platforms, the usual suspects, trapped atoms, semiconductors, photons, and superconductors. I predicted then that, within a few decades, at least one of these would evolve to the point that we would have a quantum Moore's law – namely that the number of qubits on the chip would begin doubling every few months. Joe Fitzsimons gave a talk pointing out that this had happened in 2016 – decades before I predicted it would – and that the doubling time was only six months. See Figure 6.1. At first, you might think that this is great – I'll have a quantum computer in my phone in 30 years. But the whole point of the quantum computer – the thing that makes it much more powerful than any classical computer (including my phone) – is that the processing power

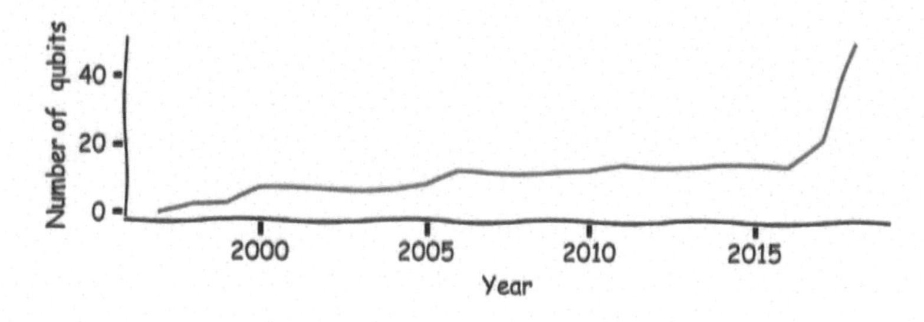

FIGURE 6.1 A plot of the number of superconducting qubits per chip as a function of year. Note the hockey stick behavior that occurs in 2016 when the number begins to start doubling every six months.

(Image courtesy of Joseph Fitzsimons.)

22 https://en.wikipedia.org/wiki/Moore%27s_law.

increases exponentially with the number of qubits. Thus, if the number of qubits is increasing at an exponential rate, the processing power is increasing at a double-exponential rate. This double-exponential behavior I also predicted in my 2013 book, and Hartmut Neven at Google observed this behavior for the first time in his quantum computer in the Spring of 2019.[23] (See Figure 6.2.)

The reason we hit quantum supremacy in 2019 is that the number of qubits started doubling every six months beginning in 2016.[24] While the

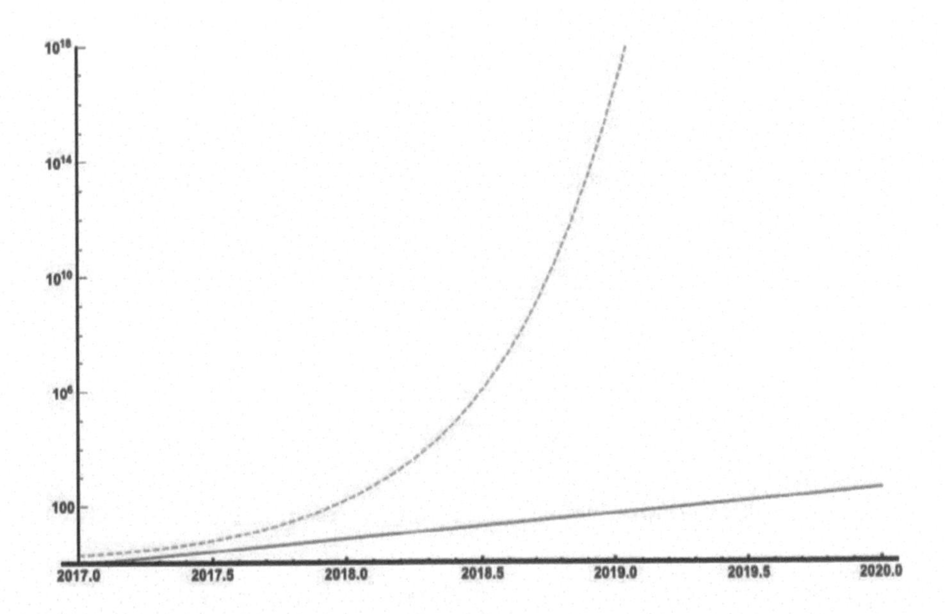

FIGURE 6.2 Here we show in a log plot the growth of the number of superconducting qubits by year (solid blue line) and the growth of the quantum processing power by year (dashed orange curve). You can see that in 2019, the number of qubits hit about 50, which is valid for Google, IonQ, and the Chinese photonic machine. The processing power is growing at a double-exponential rate, and so at this same point in time, the size of the computational space is equivalent to 100 trillion classical processors – enough to give any IBM classical supercomputer a run for its money.

23 Hartnett, Kevin. "A New Law to Describe Quantum Computing's Rise?" *Quanta Magazine*, 18 June 2019. www.quantamagazine.org/does-nevens-law-describe-quantum-computings-rise-20190618/.

24 This observation of the doubling rate was made in 2016 by Joseph Fitzsimons, www.quantumlah.org/research/group/joseph.

current number of qubits is around 50, the processing power associated with them is that of 100 trillion classical transistors. Remember, those *quantum* transistors don't all exist in our universe, but the vast majority of them live in many parallel universes – where we don't pay the electric bill.[25] This double-exponential growth is why we hit quantum supremacy in 2019 – 100 trillion transistors are around that of the world's fastest supercomputer. When Google announced it had quantum supremacy, IBM countered that Google's machine was only a bit more potent than one of IBM's most advanced classical computers – but this claim is only quibbles and bits.

Because of the six-month doubling time, by the fall of 2020, we'll have quantum computers with 200 qubits that have the processing power equivalent to a classical computer with the same number of transistors as there are atoms in 10,000 solar systems. This means that to compete with a 200-qubit quantum machine, IBM would have to make all the atoms in 10,000 solar systems into transistors and construct an interstellar classical computer that was millions of lightyears across. That's what it would take just to keep the classical computers competitive through the end of 2020. By 2028, IBM would have to turn all the atoms in the universe into a classical computer to keep up.[26]

The classical worldwide web was invented in 1990 to allow scientists at different laboratories around the world to share scientific documents. Who would have thought then that, 30 years later, the web would be used for banking, online gaming, and sharing endless videos of cats.[27]

25 For more on the many-worlds interpretation, see https://en.wikipedia.org/wiki/Many-worlds_interpretation.

26 This double-exponential growth was predicted by me in my 2013 book and observed by Neven in the Google quantum computer in the spring of 2019. John Preskill at Caltech has combined our observations into The Neven/Dowling Law: "For the largest quantum circuit that can be executed with fixed fidelity, the classical cost of the simulation is increasing doubly exponentially with time. That's really fast." Taken from Preskill's lecture "What's Next After Quantum Supremacy?" given at the QB2 conference, Practical Quantum Computing, 10–12 December 2019, held in San Jose, CA. A photo of Preskill's slide on the Neven/Dowling Law is here https://twitter.com/quantumVerd/status/1204813439601758208/photo/1. His full talk is here www.theory.caltech.edu/~preskill/talks/Preskill-Q2B-2019.pdf. What's amazing is that this talk was happening in real time while I was writing this footnote.

27 The history of the World-Wide Web can be found here https://en.wikipedia.org/wiki/World_Wide_Web. One of the oldest scientific document sharing platforms was The ArXiv, https://en.wikipedia.org/wiki/ArXiv, which began as the LANL ArXiv, as it was housed at the Los Alamos National Laboratory. The original internet domain name for the site was xxx.lanl.gov. The server back then was desktop IBM 486 PC, which had limited storage and also limited bandwidth. Back in those days, any domain name that began with "xxx" was a pornography site. The precursors to search engines such as Google Search were automated computer codes called web crawlers that

We did not know what to do with the classical internet until we started using it. Same with classical computers – the 1945 ENIAC was built to calculate Army artillery shell trajectories, but it was never used for that purpose. The first project that machine was used for was to do calculations for the design of nuclear weapons. Today my phone is trillions of times more powerful than that computer, and I mostly use it to send text messages and emails, but I can also stream videos and could play video games if I wanted to. It is impossible to predict what quantum computers hooked to a quantum internet will be capable of until we have them, and people (like my high school students) begin playing around with them to see what they can do.

We have to start training students how to program quantum computers at a very young age. In May of 2019, the vice chancellor for research at my university was invited to a White House Academic Roundtable on Quantum Information Science, where the topic of training young people in quantum technology was front and center.[28] The vice chancellor for research at MIT said, "I don't know how we can have young people working on quantum technology until they have at least had the two MIT upper division-courses in quantum mechanics." To that, my vice chancellor retorted, "Back home at LSU, Prof. Dowling has high school students programming the IBM-Q quantum computer – and I can assure you that they did not take your damn course."

A CHINESE QUANTUM-SPUTNIK MOMENT

I wrote in my 2013 book, "The future of the quantum internet is in photons, and the short-circuiting of the development of optical quantum information processors in the U.S. means that the future quantum

---

would crawl around the web looking for interesting stuff to download and catalog. The programmers who wrote the web crawler software deliberately programmed the things to eschew any domain that started with xxx to *abstain from* porn sites. (Likely this was not for puritanical reasons, but rather that there were so many such sites that all the porno data would swamp their web crawlers.) Knowing this, the arXiv designers *deliberately disguised themselves as a porn site, so the* crawlers would avoid them instead of hogging the arXiv's internet connection by downloading scientific manuscripts (instead of porn). A few years later, at a meeting at Google Headquarters, a programmer gave a lecture on how the current Google search engine works. In the Q&A, I recounted this amusing story about the arXiv domain name and asked how they handle things differently now. The Google guy retorted, "You want porn? – We'll give you porn!"

28 See www.aip.org/sites/default/files/aipcorp/images/fyi/pdf/white-house-academic-roundtable-on-qis-2019.pdf.

internet will have 'Made in China' stamped all over it."[29] I warned the U.S. government that China was pulling ahead, particularly in quantum cryptography and their long-planned quantum space mission. My government told me that "The U.S. is always at least ten years ahead of its adversaries – the Chinese quantum satellite is merely a propaganda ploy that will blow up on the launchpad."

I tried!

The Chinese launched the quantum satellite Mozi in August of 2016 – and it did not blow up – surprise! Instead the spacecraft worked extremely well – better than even the Chinese expected – and it began to take scientific data by January of 2017. In the summer of 2017, the Chinese satellite group published three astounding papers that revolutionized quantum communications forever.

The first result established two BB84-type quantum one-time pads between the satellite and two telescopes on the ground separated by 2,500 km. The two ground stations were then able to communicate with each other in absolute security using quantum keys provided by the satellite. Since then, the Chinese team has extended the range to 7,500 km between a ground station in China and a second ground station in Austria. The data rate was sufficiently high that they were able to use the quantum key to secure a video conference between the President of the Austrian Academy of Sciences in Vienna and the President of the Chinese Academy of Sciences in Beijing.[30]

The satellite carries three payloads, with the quantum crypto package being just one of them. With the second payload, which contains a bright entangled-photon source, they were able to distribute entangled photon pairs between two ground stations located 1,200 km apart. In the initial experiment, they used this setup to violate Bell's inequality, showing that quantum nonlocality holds over vast distances. The previous record, held by the Austrians, was only 143 km. Just recently, the Chinese group used their entangled source on the satellite to implement the E91 quantum cryptography protocol, which requires the Bell test as part of its workings.

Finally, the third payload on the satellite was a teleportation receiving station. Using an entangled-photon source on the ground, they launched one photon of an entangled pair up to the satellite and then

29 Dowling, Jonathan P. *Schrödinger's Killer App: Race to Build the First Quantum Computer*, 173. Taylor & Francis, Boca Raton, FL, 2013.

30 See https://en.wikipedia.org/wiki/Quantum_Experiments_at_Space_Scale#Mission.

teleported the quantum state of a third photon from the ground to the photon on its way to the satellite – a distance of 1,400 km. The previous record, again held by the Austrians, was a mere 143 km.

In spite of my repeated warnings over many years, these three experiments caught the U.S. government completely off guard. Remember, we expected the rocket to blow up! The Mozi satellite is an interesting object-lesson for the Chinese approach to their quantum space program. The primary payload was the crypto payload – that's how they got their government to pay the bill – by touting absolute security against attacks from classical and quantum computers alike. The sole purpose of the Bell-test experiment was to test the foundations of quantum mechanics – a field of interest that the director of the satellite program, Jian-Wei Pan, has been passionate about all of his life.[31] The teleportation experiment sits somewhere between foundation and practice; while it too tests the laws of quantum theory, teleportation is a primitive of any quantum network, as we have discussed in the previous chapter.

It is of interest to note that Pan's mentor (and occasional competitor) Anton Zeilinger[32] had a similar scheme to test quantum theory from space that he had proposed a decade ago to the European Space Agency.[33] The project was not funded. The genius of the Chinese program was to sell their satellite to their government primarily as a state security tool and then add a couple of hitchhiking payloads to test the quantum theory. The Mozi satellite is coming to the end of its lifespan. As I write this, the single-photon detectors on the satellite have become almost useless due to space radiation damage. However, the entangled-photon source seems to be still working fine, so they can continue doing experiments where they send photons down, like crypto and Bell testing, but they can no longer do experiments like teleportation where photons are sent up to the satellite.

31 See https://en.wikipedia.org/wiki/Pan_Jianwei.

32 See https://en.wikipedia.org/wiki/Anton_Zeilinger.

33 Ursin, Rupert , Thomas Jennewein, Johannes Kofler, Josep M. Perdigues, Luigi Cacciapuoti, Clovis J. de Matos, Markus Aspelmeyer, Alejandra Valencia, Thomas Scheidl, Alessandro Fedrizzi, Antonio Acin, Cesare Barbieri, Giuseppe Bianco, Caslav Brukner, Jose Capmany, Sergio Cova, Dirk Giggenbach, Walter Leeb, Robert H. Hadfield, Raymond Laflamme, Norbert Lutkenhaus, Gerard Milburn, Momtchil Peev, Timothy Ralph, John Rarity, Renato Renner, Etienne Samain, Nikolaos Solomos, Wolfgang Tittel, Juan P. Torres, Morio Toyoshima, Arturo Ortigosa-Blanch, Valerio Pruneri, Paolo Villoresi, Ian Walmsley, Gregor Weihs, Harald Weinfurter, Marek Zukowski, Anton Zeilinger. *Space-QUEST: Experiments with Quantum Entanglement in Space.* ArXiv, 2008. https://arxiv.org/abs/0806.0945.

There are plans on the books to use what they learned with Mozi I in the design of Mozi II, which will have a brighter entangled photon source and better shielding of the detectors. In addition, the Chinese have plans to launch about ten nanosatellites in the near future. The nanosatellite can be about the size of a cell phone with solar panels sticking out of it – their sole payload will be designed to distribute the BB84 quantum key to the different ground stations. In this way the Chinese government intends to eventually secure the entire country using their satellites in orbit and quantum networks on the ground. History changed in 1957 when the Soviets launched Sputnik – the first orbital spacecraft. History changed again in 2016 when the Chinese launched Mozi. Then as now, the U.S. thought they were at least ten years ahead of their competitors in the space race. Sputnik gave the U.S. a well-deserved boot in the buttocks – immediately after its launch the U.S. Department of Defense responded to the political fallout by approving funding for the launch of the first U.S. orbital satellite Explorer. Sputnik also led directly to the founding of NASA, the subsequent race to the moon, and the overhaul of the U.S. education system in math and science.

Today, after Mozi, it's like Déjà vu all over again!

NATIONAL QUANTUM INITIATIVE ACT

Just like Sputnik led to the founding of NASA, Mozi led to the founding of the new U.S. National Quantum Initiative Act.[34] In June of 2017, right when the data from Mozi was made public, it became immediately clear the satellite was not a hoax. The Science Committee of the U.S. House of Representatives approached the U.S. National Photonics Initiative (NPI)[35] and asked for a white paper outlining a U.S. response to the Chinese breakthrough.[36] The NPI responded with an 11-page document titled "Call For A National Quantum Initiative: Proposal to address the urgent need to develop the information infrastructure of tomorrow." This document served as a primer for the National Quantum Initiative Act, signed off by the President in December of 2018. That is amazingly quick for the U.S. government. In only 18 months,

34 See www.aip.org/fyi/2019/national-quantum-initiative-signed-law.

35 Why did they approach the National Photonics Initiative? The satellite was transmitting and receiving photons. See www.lightourfuture.org/home/.

36 A copy of the NPI white paper is here www.lightourfuture.org/getattachment/Home/About-NPI/Resources/NPI-Recommendations-to-HSC-for-National-Quantum-Initiative-062217.pdf.

they went from "the Chinese quantum satellite is a hoax" to the National Quantum Initiative Act.

The NPI document declared that the three pillars of the National Quantum Initiative are:

1. quantum-limited and quantum-enhanced sensors of force fields and time,
2. optical photonic quantum communication networks,
3. quantum computers.

All three of these topics fall under the rubric of quantum information science, which turned out to be the topic of the final version of the National Quantum Initiative Act. It may seem a bit odd that quantum sensors are considered part of quantum information science, but there is a reason for that. Way back in February of 1998, I gave the first public talk on quantum sensors, "From Quantum Computers to Quantum Gyroscopes," at the NASA International Conference on Quantum Computing & Quantum Communications, in Palm Springs, California. I had been scheduled to give a talk on another topic, and my time slot was in the morning on Friday of the last day. I never gave my planned talk. Instead, after listening to the other presentations all week, I had one of my most profound ideas – a quantum sensor is a special-purpose quantum computer! A quantum computer is a machine that exploits unreality, uncertainty, and entanglement to solve mathematical problems much faster than a classical computer can.

My recent work had shown that a quantum sensor was also a machine that exploits these three pillars of quantum technology to measure things much more precisely than a classical sensor can. I made the case that advancements in quantum computing would directly lead to revolutions in quantum sensing and vice versa. I tossed out my original presentation and at the hotel bar the night before I was scheduled to speak. I got out a stack of blank transparencies and my special Lumocolor™ pens and wrote an entirely new talk on quantum sensors meet quantum computers. This talk turned out to be a game-changer – not only for the field, but for me personally.[37] Ever since that day, quantum sensors and quantum

37 At the time of that conference, I was a research physicist at an Army Laboratory in Huntsville, Alabama, but I was looking for a new job where I could spread my wings. After the talk in Palm Spring, I was mobbed by a bunch of the NASA conference organizers. They were very excited

computers have been tied together at the hips. During the questions after my talk, Jeff Kimble retorted, "You can't just take ideas like quantum error correction and apply them to quantum sensors!" Today – 21 years later – there are several multi-million dollar government-funded research projects devoted to doing precisely that.

Motivated by the NPI white paper, a new document, prepared by the National Science and Technology Council, appeared in September of 2018. It is titled the "National Strategic Overview for Quantum Information Science."[38] This document greatly expanded upon the material in the NPI report, and it became the template for the National Quantum Initiative Act. The focus of the overview remained on quantum information science, as defined above. What was new is that it also contained a strong focus on education and training of the future quantum information workforce. Right now, there is a terrible shortage of young people trained in quantum technology – at a time when the job market for such persons is exploding at an exponential rate. The few who are trained in the area are often lured first into high-paying jobs in industry or start-up companies. Not just junior people, tenured professors are taking leave of their posts – and of their senses – and founding new start-up companies.

---

since quantum computing and quantum communications had been a hard sell to the NASA program managers – but everybody at NASA likes a better sensor. They invited me to come to the NASA Jet Propulsion Lab in March of 1998 to re-present my Palm Springs lecture, and they lured me with vague promises of funding for a postdoctoral intern of my own at the Army lab. I gave my talk on a Thursday afternoon and was to meet with JPL managers on Friday. Thursday evening, some old college friends of mine drove into town, and we went out for Mexican food and lots of margaritas to celebrate my lecture. The next day I was a bit hungover but soldiered through the morning meetings fine. For lunch, I met with Colin Williams, who was running the nascent JPL quantum computing activity at the time. We had many interesting discussions, and then after lunch, we had an appointment to meet with the director of the JPL high-performance computing group. I suspected it was he who would be offering me funding for my intern. But as Colin and I boarded the elevator up to the director's office, Colin suddenly panicked and turned to me, "I completely forgot! The director is going to offer you a job here at JPL! I was supposed to discuss the details with you at lunch – so play along for now, and I can fill you in later." In that 20-second elevator ride – my entire life flashed before my eyes. We entered the director's office, and he said, "Are you alright? Your eyes are all red." It was from my hangover, but I replied, "Oh – no worries – it's just allergies." Nodding, the director continued, "I assume Colin filled you in on all the details about the job offer?" Colin turned to me – nodding nervously. "Oh – yes!" I said, "It is exciting news!" I got the first written offer in about April and then negotiated the salary up because living in Pasadena was twice as expensive as in Huntsville. In August, while at my family reunion, I finished the negotiations over the phone at my parents' house and then announced to my family, "I got the job at JPL!" My father surprised us all by pulling out a couple of bottles of champagne that he had secretly stowed away in the fridge for just such an event. Dad worked for NASA on the lunar program in the 1970s and was very happy to see me join JPL.

38 "National Strategic Overview for Quantum Information Science." www.whitehouse.gov/wp-content/uploads/2018/09/National-Strategic-Overview-for-Quantum-Information-Science.pdf.

The next most lucrative market is in the government labs, where nearly every single government agency now has a windfall of new cash in quantum information science. Still they can't hire people fast enough to do the work. A defense contractor contacted me some months ago about hiring one of my students. The only qualification was that she should know the difference between a qubit and a bit. They just hired one of my undergrads even before she graduates in May. My last three Ph.D. students, who graduated in the previous year, all took high-paying jobs at government labs – not as high-paying as in industry, but the job security is better.

Academia, which has the lowest pay and the highest level of job security (once you get tenure), is hurting the most. My colleagues find it nearly impossible to hire postdoctoral interns – as they are all going into industry

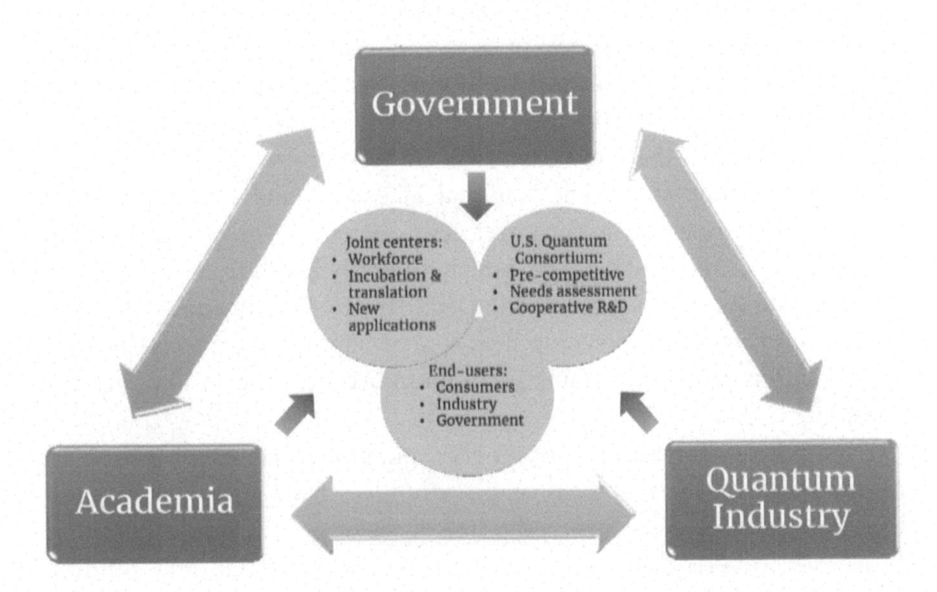

FIGURE 6.3 The graphic is taken from the "National Strategic Overview for Quantum Information Science," and it shows the interconnection between academia, industry, and government labs that is promoted by the National Quantum Initiative Act.[39]

39 This figure was produced by U.S. Government employees as part of their official duties and is therefore not subject to copyright www.whitehouse.gov/wp-content/uploads/2018/09/National-Strategic-Overview-for-Quantum-Information-Science.pdf.

or government. Postdocs are typically on the first rung of the ladder to a tenured academic job – but hardly anyone is climbing that ladder in quantum technology. We won't have people at universities to train the next generation of quantum technologists. To ramp up the training in quantum technology, the report recommends setting up master's degrees at universities and training certification courses at technical colleges. I wonder who they will find to teach those classes? Once somebody knows enough quantum technology to teach such a course, they could easily take a much higher-paying job in the private sector. The strategic overview stresses the importance of tight collaborations between industry, academia, and government labs to address such problems. See Figure 6.3.

The strategic planning document was used to write the text of the National Quantum Initiative Act. The act passed both the House and the Senate in the fall of 2018, with unusual bipartisan support, and was signed into law by the President on 21 December 2018.[40] The bill earmarked $1.2 billion for new investments in quantum technology. Most of the funds were allocated to the Department of Energy, then next most went to the National Science Foundation, and the least to the Department of Commerce. NIST is in the Department of Commerce, and it is the agency running the entire show. The National Science Foundation is a natural home for such research training and development.[41] The odd agency out was the Department of Energy, which hitherto had little experience in quantum information science.[42] The Department of Energy does run most of the government's supercomputers – so I suppose everybody thought that quantum computers should go to them because they are super-duper computers.

I attended a Department of Energy quantum planning meeting in November of 2018, and most of the attendees from that agency acted like classical deer staring into the headlight of an oncoming quantum train. Nevertheless, the National Quantum Initiative allocates funding for up to five new Department of Energy (DoE) centers in quantum technology, funded at $25 million per center over five years.[43] Less

---

40 It was the last piece of legislation the President signed before he shut down the government.

41 A summary of the National Science Foundation initiatives in quantum technology is here www.nsf.gov/mps/quantum/quantum_research_at_nsf.jsp.

42 A summary of recent DoE activities in quantum information science is here www.energy.gov/articles/department-energy-announces-218-million-quantum-information-science.

43 The full text of the National Quantum Initiative may be found here www.congress.gov/bill/115th-congress/house-bill/6227/text.

well-funded centers will be organized by the National Science Foundation (NSF). The proposals for the NSF centers are currently (7 December 2019) under review and calls for the DoE center proposals are beginning to come out right about now. About one-third of the funds for these centers is to be dedicated to training and education, with the other two-thirds allocated to basic research and the development of laboratory infrastructure. Vast amounts of money are beginning to flow. I hope the coordination is in place to leverage that money – or else it will look like the government is throwing soggy stacks of hundred-dollar bills at the walls to see where they stick.

The elephant in the room, which I have not discussed at all, is that the NQI makes no mention whatsoever of the Department of Defense (DoD). That department has its own separate budget for quantum technology that is distinct from the NQI. The DoD has traditionally been a stalwart supporter of quantum information science, particularly in such agencies as IARPA, DARPA, as well as the Army, Navy, and Air Force. The base quantum funding in these agencies will see an increase starting in 2020.[44] The breakdown of the total budget for quantum information sciences in the DoD is not given. Still the DoD is focusing on the development of quantum sensors, which includes quantum atomic clocks, while continuing to contribute to quantum computing and communications across all of the DoD agencies.

If we look at the three pillars of the National Quantum Initiative, we see that we have quantum sensors, quantum networks, and quantum computers. Hooking up distributed quantum computers over a quantum network is a no brainer – that is what the ARPANET did with classical computers decades ago. However, besides, there is a great deal of interest by the DoD and the DoE in hooking up quantum sensors over quantum networks. We'll discuss this more below, but remember the takeaway message from my 1998 speech in Palm Springs – a quantum sensor is merely a quantum computer that has been reprogrammed to make precision measurements rather than solve hard math problems. If you can hook quantum computers together on the quantum internet, then you automatically can hook quantum sensors together on the same network.

44 "FY20 Budget Request: DOD Science and Technology." *American Institute of Physics* (28 March 2019). www.aip.org/fyi/2019/fy20-budget-request-dod-science-and-technology.

QUANTUM INTERNET 1.0 – ALL THAT IT'S CRACKED UP TO BE?

The quantum internet will roll out in three stages. Quantum internet 1.0 is already here! This first stage is a network that's sole purpose is to distribute the quantum key to users in order to secure their data. The network will have ground-to-ground fiber links, ground-to-ground free-space links, and ground-to-ground links via satellite. This first version of the quantum internet has been plodding along in fits and starts since about 1994. In that year, the same year as Shor's quantum cracking algorithm appeared, a group in the U.K. announced BB84 key distribution over a distance of 10 km.[45] This result was announced by Paul D. Townsend at the 1994 International Quantum Electronics Conference held in Anaheim, California, United States, 8–13 May 1994.[46] I was there in the room when Paul gave the talk - and I was dumbfounded. Previous experiments had been carried out over meters - and I thought the whole quantum cryptography thing was a gimmick that would never be useful. But 100 km? - that's the range of a wide area network - quantum cryptography is useful! Upon returning to my job at U.S. Army Missile Command, I wrote up all that I had learned about quantum cryptography and faxed it off to the Army Research Office, which I implored to take seriously. They did and asked me and two of my colleagues, Henry Everitt and Charles Bowden, to organize a workshop on this topic for the early spring of 1995.[47] This was the first government workshop on quantum cryptography.

45 Townsend, P.D. "Secure Key Distribution System based on Quantum Cryptography." *Electronics Letters* 809 (12 May 1994): 809–811. https://ieeexplore.ieee.org/document/289239.

46 Conferences were often held in Anaheim because Disneyland was right across the street. The organizers encouraged the attendees to bring their kids who could frolic at the amusement park while the parents were in the meetings. The conference was at the Anaheim Convention Center, and the conference hotel was the nearby and very expensive Hilton, which most of U.S. poor scientists could not afford. In May of 1994, I was still a postdoctoral intern on a tight budget. I asked around, and a colleague told me about this nice clean, cheap motel right next to the Hilton called the Peter Pan Motor Lodge. That's where my fellow interns and I stayed, and that motel became the best-kept secret for meetings in Anaheim. I had a Nobel laureate call me up and ask about it. Imagine this dated motel decorated with cartoons from the Disney film Peter Pan filled with physicists bumbling around and scribbling on laboratory notebooks while spouting what appeared to be nonsense.

47 As recounted in my previous book, after hearing Artur Ekert's talk on Shor's algorithm in July, I sent a second fax on quantum computing to the Army Research Office, and the first Department of Defense workshop on quantum computing and quantum cryptography was born. It was held in Tucson in February of 1995.

Since the 1994 breakthrough, a handful of small quantum crypto local area networks have sprung up over the years since, but with many of them eventually withering away.[48] Most of these networks were proof-of-principle implementations but were never put to any practical use. The DARPA Quantum Network (2004–2007) had ten nodes distributed around Boston. Again, this was a testbed with no practical use. Another similar network, which operated in the same time period (but had no name), had nodes at secure government facilities distributed around Washington DC.[49] In 2004 the quantum local area network (QLAN) of Zeilinger's group in Vienna secured banking data with quantum key distributed under the Danube river, but this was a one-off stunt. However, by 2008, this network had evolved into something that linked Vienna and a few nearby cities, but it was mothballed in 2010.[50] Similarly, Nicholas Gisin and his team ran the SwissQuantum network project (2009–2011) that had several nodes in greater Geneva, Switzerland. A precursor to this network was used to secure some of the vote tallies in the 2007 Swiss federal election, but again this was another one-off stunt.

The first truly practical and continuously used quantum crypto network was once again built and operated by the Chinese. In 2017, a 2,000 km fiber-trunk line carrying BB84 key had been installed all the way from Shanghai to Beijing along the route of a high-speed train line. The trunk-line connects BB84-based QLANs in Shanghai, Hefei, Jinan, and Beijing.

---

48 Quantum Key Distribution Implementations, https://en.wikipedia.org/wiki/Quantum_key_distribution#Implementations.

49 In early 2004, when I still worked at the NASA Jet Propulsion Laboratory, a DARPA program manager called me up and asked to come visit me in my office in Pasadena. (This was a bizarre request, as usually DARPA program managers call you up and order you to visit them in their office in Washington, DC.) I was happy to comply, thinking that this guy might want to fund my research. He shows up in my office, and begins talking about the DARPA Quantum Network in Boston (that he funded). Then he asked if I knew about the one around Washington, DC. I said that I could neither confirm nor deny that I had heard of it. That was enough – and he pops the question – "What I am looking for is a satellite we can use as a quantum relay to exchange key between the Boston network and the DC network." Puzzled by how I could possibly help with that, I replied, "How can I possibly help with that?" He responded, "I was wondering if you could talk to your folks at NASA about lending me a NASA satellite to do the job." I was astonished but secretly pleased that a DARPA program manager thought I had some joystick in my office that would let me send orbiting NASA satellites careering about at my will. I told the DARPA guy that I would talk with the JPL satellite laser communications folks, but not to get his hopes up. Then I asked the elephant-in-the-room question, "You work for the Department of Defense! You have plenty of your own satellites up there – why don't you use one of yours?" He paused for a moment and then (in a low voice) said, "I tried – but they won't talk to me. They think this quantum cryptography is useless and a big waste of time."

50 Secure Communication based on Quantum Cryptography https://en.wikipedia.org/wiki/Secure_Communication_based_on_Quantum_Cryptography#Brief_architecture_of_SECOQC_network.

The system went out of the test phase late in 2017 and has been continuously operating ever since – distributing key that is used to secure financial and government data transmissions. Since quantum repeaters don't exist yet, they rely on having secure relay nodes on the network spaced 10–50 km apart – typically located in a secure room at each train station. I have seen this system in operation from the control room in Shanghai, and it is quite a sight to behold. In the control room, there is a giant screen displaying all the links and nodes in the networks with flashing lights telling you what the key rate is at any node is in real time. This Chinese network is not a testbed or a publicity stunt – it is the real deal – and it will go down in history as the fully operational quantum network 1.0. It is not practical to roll out something like this all across China via landlines, at least not without a quantum repeater, and so the Chinese intend to connect the further-flung QLANs around the country to this primary network via quantum satellite links.

No other country has *anything* like this.

QUANTUM INTERNET 2.0 – RAGE OF ENTANGLEMENT

Before post-quantum cryptography came along, I thought Quantum Internet 1.0 would naturally evolve into 2.0. Now that there is a post-quantum

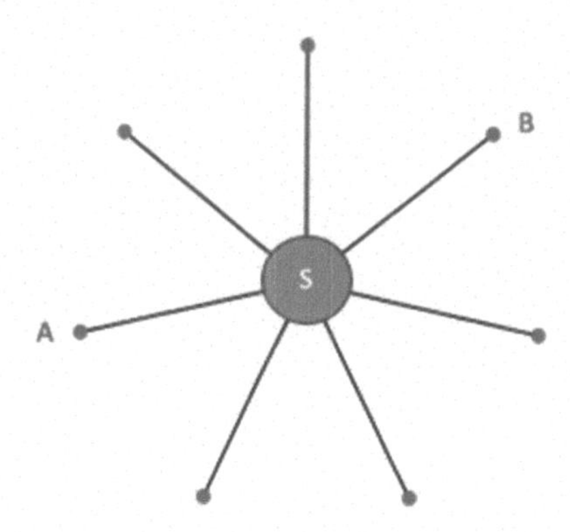

FIGURE 6.4 The star-graph network topology. The users are the red dots on the outer edges of the graph, and the green dot in the center is the switching and entanglement distribution hub.[51]

51 Star graph https://commons.wikimedia.org/wiki/File:Star_network_7.svg. This file is licensed under the Creative Commons Attribution-Share Alike 3.0 Unported license.

offramp – who knows? If 2.0 does appear, I again expect it to first happen in China, where some of the critical elements have already been tested. As the quantum networks mature, eventually, the quantum memories and the quantum repeaters will come online, and the purpose of the networks will evolve from distributing quantum key to distributing quantum entanglement. Once you have entanglement, you can secure it with the E91 key and do other things besides.

A hundred years ago, we had a classical network called the telephone system. The local-area networks consisted of a star-graph topology, as shown in Figure 6.4.

If you watch primordial films from the 1930s, the telephones at the user nodes (outer red dots) were equipped with a mouthpiece, an earpiece, and a bell.[52] (See Figure 6.5.) The telephone company was smart in that it wanted to reach customers who might not yet have electricity or even running water. The central hub provided common electrical power to the nodes. Those electrical links were used to transmit analog voice signals and also to ring the bells on the phones.[53] Before the invention of the dial telephone, the user Alice (A) would pick up the receiver and turn a crank that would ring a bell at the central switchboard (S). When a switchboard operator answered, Alice would state that she would like to talk to Bob (B). The operator would then turn a crank to ring the bell on Bob's phone. Bob would pick up his receiver, and the operator would tell him that he was getting a call from Alice. The operator would then take the end of the wire running from Alice's house and plug it into a jack to connect it to the wire that ran to Bob's house. Then the two of them could speak. The phone company provided two critical services – electrical power from the source to all of the user nodes and a switching capability at the source.

Let's now use this analogy to construct a quantum-star network. At the switching station S, there are now lots of sources capable of generating lots of entangled photon pairs. (Since a qubit is a quantum bit, an ebit is defined to be an entangled pair of qubits.) There are also many exquisite quantum memories and machines for making perfect

52 History of the telephone https://en.m.wikipedia.org/wiki/History_of_the_telephone.

53 In my home, I still keep one of these old-fashioned landlines and a phone that requires no electricity other than that from the phone line. Here in Louisiana, when the hurricanes hit, the first thing to go out is the power grid, and then all the cell-phone towers fail. However, the power provided by the phone company often lasts for days after the hurricane hits, and I can use the landline to call for help or assure friends and family that I'm okay.

FIGURE 6.5 An 1896 telephone from Sweden. The mouthpiece and receiver are on the handset in the cradle on top, and the crank is visible on the side. For this model, the bell was in a box under the table.[54]

Bell measurements – in other words, lots of entanglement swappers. In the quantum internet 2.0 the memories, ebit sources, and swappers all take up one large lab bench each. Even for a small quantum-star network, the switching station S will need to be the size of a warehouse, and for a more extensive system, it could take up an entire city block. Even in 1999, the New York City telephone switching station was a towering, windowless skyscraper.[55]

The users, Alice and Bob, will be located in large companies where there will be an entire room or floor of a building devoted to receiving and storing shares of ebits. The users will have, at a minimum, a large quantum memory and a transducer that allows for the conversion of incoming photonic qubits into solid-state memory qubits. In quantum internet 2.0, only large companies, universities, and government labs can afford this kind of setup. The network is analogous to the ARPA-NET, where only such places could provide the big computers that the webbing hooked together.

---

54 This work is in the public domain in its country of origin and other countries and areas where the copyright term is the author's life plus 70 years or fewer, which includes the graphic taken from https://commons.wikimedia.org/wiki/File:1896_telephone.jpg.

55 The AT&T Long Lines Building https://en.wikipedia.org/wiki/33_Thomas_Street.

In the case of this quantum network, all the nodes in the star in Figure 6.4 are connected to the central station S by fiber or free-space links, but none of the nodes are connected to each other directly. Let's call the company that runs this network American Qubit and Ebit (AQ&E). What they are selling is entanglement. Assume for the moment that the source and the users all have perfect memories that can store qubits - without degradation or error - for arbitrarily long times. We will also assume that the ebits generated at the source are darn good and that when one qubit (that makes up half of the ebit) is transmitted from the source to the user, there is no loss or degradation of the qubit. (If there is loss, then that decreases the rate at which the users get their qubits from the source. If there is decoherence, then that can be compensated by carrying out entanglement distillation or purification – as described in the previous chapter.)

Alice and Bob purchase data plans from AQ&E. The company is selling ebits. Let's suppose Alice and Bob both have unlimited data plans. Alice uses the classical internet (which is a separate system in quantum internet 2.0) and places an order for 24 billion entangled qubytes. Each qubyte contains eight qubits. She is thence ordering 24 GQ quantumly instead of 24 GB classically. That works out to about 200 billion qubits. AQ&E then directs the source to produce 200 billion *ebits*, and send one qubit in each ebit to Alice and to store the other qubit in the ebit in a local quantum memory at the switching station. When Alice's qubits arrive, she stores them in her local quantum memory. Bob carries out a similar procedure. See Figures 6.6 and 6.7.

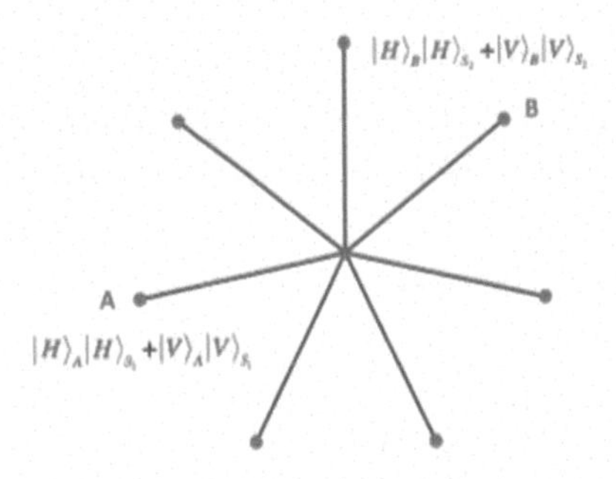

FIGURE 6.6 Alice and Bob instruct the central switching agency to send them each half of an ebit in the form of entangled photon pairs. One photon in one pair is emitted to Alice, and the other kept by the switching station. Another photon in a second pair is sent to Bob, and its entangled partner is also held at the switch.

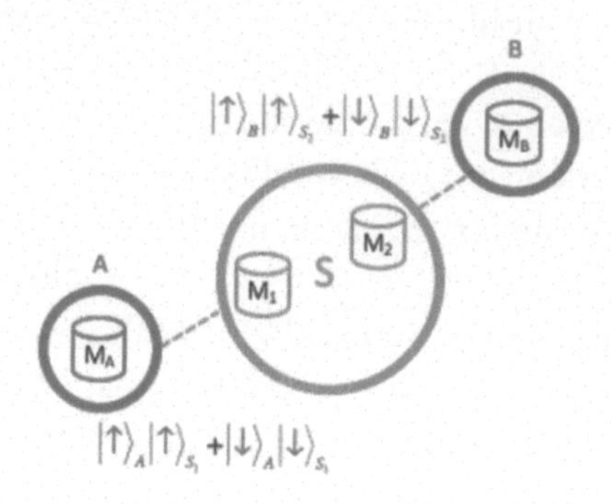

FIGURE 6.7 After the entanglement is distributed via photons (Figure 6.6), Alice transduces her photons into a spin qubit in her quantum memory M_A. Bob does the same with his qubit in his memory M_B. The central switching station has two memories of its own – M_1 for the other half of Alice's ebit and M_2 for the other half of Bob's. As shown, at this stage, there are two ebits stored in four memories – the Alice-Switchboard ebit and the Bob-Switchboard ebit. The next step is for the switchboard to carry out entanglement swapping on the two qubits it has in M_1 and M_2.

Alice's data plan includes the cost of generating the ebit and storing half of it in her company's memory. This protocol is something like cloud computing, but instead of all of your data residing in the cloud, half is in the cloud, and the other half is on your local memory. Bob has a similar plan, and, as per Figure 6.4, all seven users at the outer nodes share billions of ebits with the company's central switching station S. We want the system to enable any of the seven users to share lots of entanglement with any of the other seven users.

For concreteness, let's suppose Alice wants to share many ebits with Bob. She contacts AQ&E over the phone, or the classical internet, and places an order to share entanglement between each of the 200 billion qubytes she has in her memory with each of the 200 billion qubits that Bob has in his. The company rings up Bob to confirm the transaction and, when both users agree, the company instructs the switching station S to carry out 200 billion entanglement swaps. As we recall, the swap operation destroys any information in the qubits at the switching station, and in the end, Alice shares 200 billion ebits with Bob, and the station has nothing. At the end of the swaps, Alice and Bob share ebits – with the switchboard entirely out of the picture (Figure 6.8).

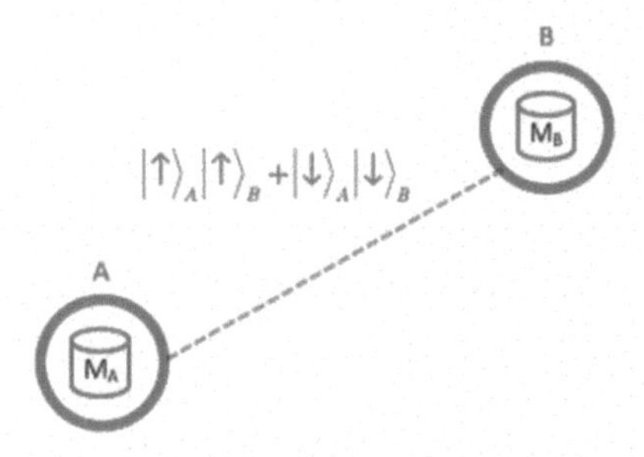

FIGURE 6.8 After the switchboard completes the entanglement swapping operation, it's two qubits are destroyed, leaving only a single ebit shared between Alice and Bob. Alice and Bob can then use this ebit for cryptography, distributed quantum computing, or teleportation, without any further dealings with the switchboard.

To make sure that the station has not been hacked or spoofed, Alice and Bob select some of their shared ebits at random and run a Bell test on them. If the Bell test succeeds, then they are sure that the station knows nothing. If the Bell test fails, then that means either the station has been compromised, or there are eavesdroppers on the transmission lines. In this way, they verify that they have entanglement – independently of whatever was going on in the switching station. This setup is called a device-independent protocol, since the security can be proven regardless of the details of the machine producing, storing, and swapping the entanglement. After the Bell tests, Alice and Bob are sure they share lots of ebits with each other and then are free to do with these ebits whatever they'd like – quantum cryptography, quantum teleportation, or link their two quantum computers or quantum sensors together. The company provides entanglement and switching that enables all of this – in analogy to AT&T that provides electrical power and switching.

Note that, after Alice and Bob receive their entanglement, somebody could blow up the switching station, and it would not affect their ability to do any of these things. They would still need access to the classical internet or phone system, which is independent of the quantum switchboard, to implement some of these things like teleportation, but once they have their entanglement, they are free of the switching station until they run out of it. This scenario is in contrast to the classical situation with AT&T, where all the communications go through the switching station, and nothing can be done without it. Perhaps a better analogy is that, once you charge-up your phone, you can survive for a couple of days without the power grid.

These star networks will at first have a short range where the nodes are separated from the switch by 10–50 km, with no repeaters, and perhaps a bit bigger if we allow a few repeaters on each link connecting each node to the switch. This is the quantum version of the AT&T local phone call – long-distance calls are much more expensive. To extend the distance, we imagine that each city has its own star network, and the goal then is to hook the stars together over long-distance trunk lines allowing the switchboards at the center of each star to share entanglement with switchboards with its nearest neighbor stars. The trunk lines will have to go distances of hundreds to thousands of kilometers, and so (if they are fiber-based), they will require lots of very good and closely spaced quantum repeaters. Just as making long-distance calls used to be very expensive – long-distance entanglement sharing will also be costly. An extensive quantum network will then look like that in Figure 6.9.

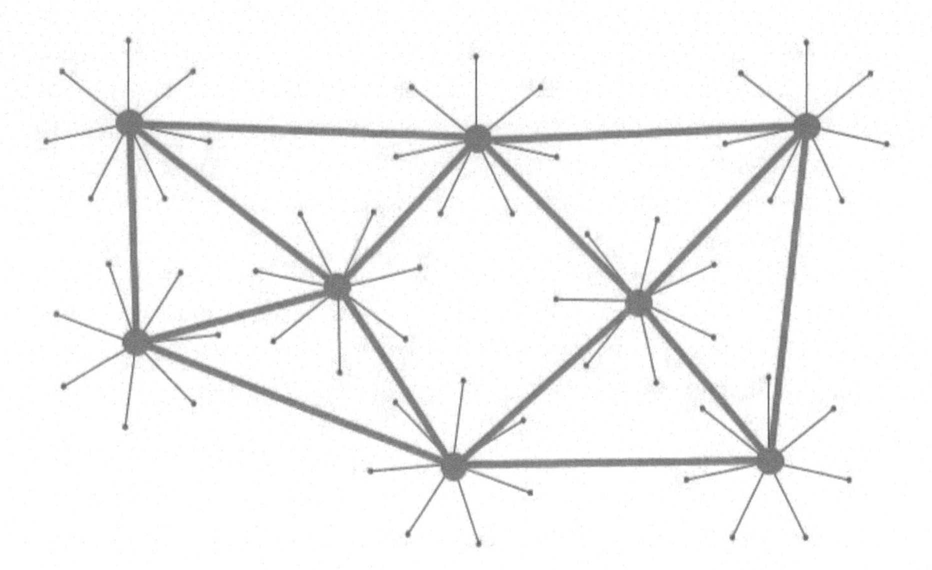

FIGURE 6.9 Quantum local-area networks (QLANs) in star formation are connected via entanglement distributed over the thick green trunk lines. In this way, users at nodes on different stars may share entanglement over vast distances.

The nodes in each star are no more than 100 km apart (without repeaters), but the distance between stars can be 1,000 km. If the trunk lines are ground-based fiber quantum communication channels, losses will require very good quantum repeaters spaced quite closely – every 10 km or so – along the trunk lines. With current technology, every repeater station would be the size of a large room. An alternative approach is to connect the stars via satellites that distribute quantum entanglement to pairs of ground stations from space. Our group's recent work shows that satellite links are better than ground links, with current technology, in that the satellites can distribute entanglement to pairs of QLANs at a much higher rate than doing it over the quantum landlines. (See Figure 6.10.)

In the figure, we show a constellation of satellites distributing entangled photon pairs to widely separated ground stations. The record separation for this type of experiment, carried out by the Chinese satellite Mozi, is about 1,200 km. Once the photons arrive at the ground stations, they are transferred into quantum memories. Upon repeated swappings, the entanglement can then be shared by any two users in the widely separated QLANS. In this way, we will establish a global network of quantum entanglement distribution.

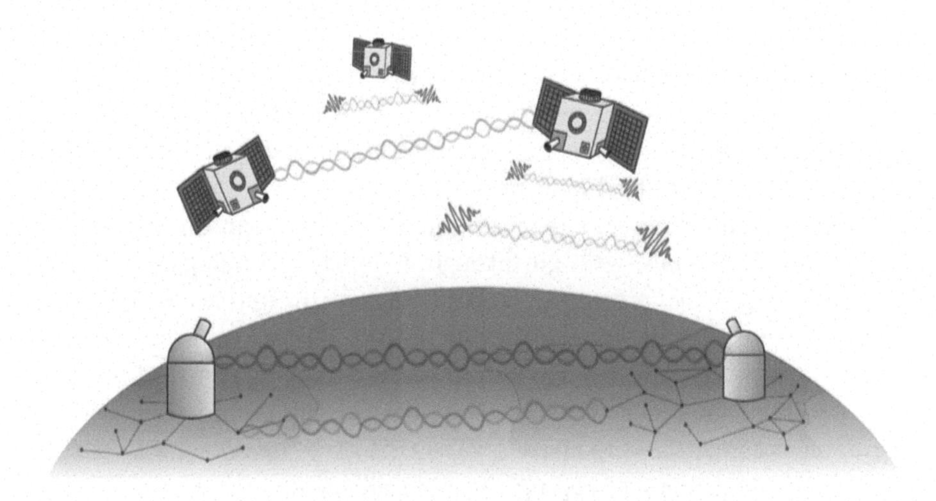

FIGURE 6.10 Two intercontinental QLANS linked together by entanglement distributing satellites. Each satellite emits trillions of pairs of entangled photons per second. The two photons arrive at the two ground stations as shown, where the user transduces the entanglement into quantum memories.

QUANTUM INTERNET 3.0 – RISE OF THE UBIQUITOUS QUANTUM REPEATERS

Recall that, in the 1970s, the telephone system in the U.S. ran over a network of copper wires. These wires could not carry much data very far, and so to supplement those landlines – global communication satellites were used. That is the reason that some of my long-distance calls had a weird delay when I was talking to my grandmother – the signals were bouncing up to satellites and back down. Nowadays, you never hear this delay as all the copper wires have been replaced by optical fiber, which can carry far more data a much greater distance. We no longer use satellites for common-place long-distance phone calls, except in places where you can't run a fiber – say out to a cruise ship. Even then, the ships have to use specialized transmitters and receivers to communicate with the satellite.

What allowed us to make that upgrade? First, companies like Corning made the fibers very long and with extremely low loss. Second, the invention of the classical repeater – a laser-like device integrated into the fiber – allowed you to put repeaters every few hundred meters. The classical-network breakthrough came when we made the repeaters small, cheap, and easy to integrate all over the network. For example, there are classical repeaters every few hundred meters on fibers in undersea communication cables that span the entire Atlantic Ocean. The repeaters are so small and simple that they are bundled together in the fiber cable along with a battery to provide them with power. They use so little power that the battery lasts the lifetime of the cable itself. By the year 2000, nearly all satellite communication links had been phased out in favor of the optical fiber network.[56]

Someday something similar will happen with the quantum repeaters. When that day comes, we'll go through a similar technological revolution. The repeaters, memories, and entanglement sources will fit on a chip in a box that sits next to your computer. We will call it a q-router instead of a router. We are working on these things right now. I'm part of a program, funded by the U.S. National Science Foundation, where we develop computer-chip-sized devices that have the entangled source and the entanglement swapping machine all on the same chip. If we don't succeed, somebody someday soon will, and we'll be able to replace all those giant switches and sources at the switching stations with

56 Transatlantic Communications: Current Technology https://en.wikipedia.org/wiki/Transatlantic_communications_cable#Current_technology.

gizmos the size of a matchbox that are powered by a couple of AAA batteries. For a quantum repeater, though, we need a quantum memory about the same size. Investments in quantum memories continue to grow, and soon we could have something like our diamond memory, discussed in the previous chapter, also on the same chip as the source and the swapper. Once this breakthrough occurs – and occur it will! – cheap and easy-to-make entanglement will be ubiquitous. Everybody will have these entanglement-distribution nodes in their homes and phones. Quantum repeaters will be so small that we can pack them every few hundred meters on undersea quantum optical-fiber communication cables. The third quantum-internet revolution will give unlimited entanglement to the people.

What will the internet 3.0 look like? People are thinking about that now. In particular, my students are thinking about this. Consider the quantum network in Figure 6.11. The network no longer has the star structure of

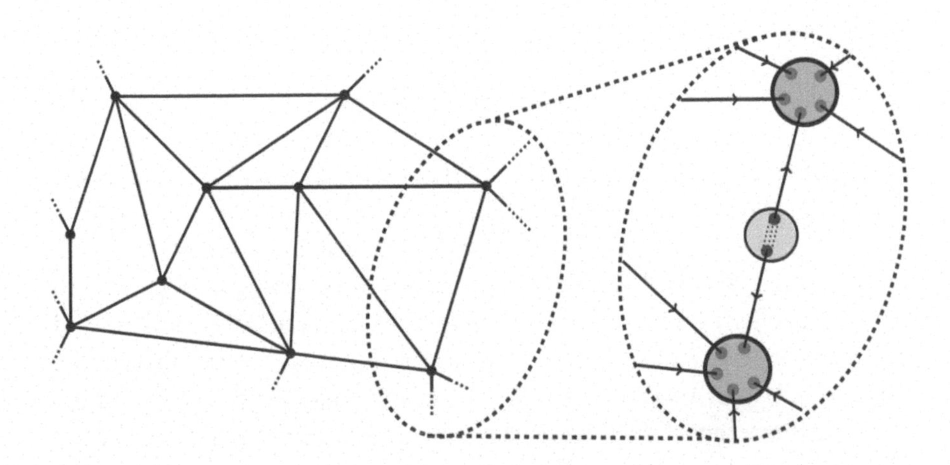

FIGURE 6.11 Quantum internet 3.0. Left: the network consists of user nodes (black dots) connected by fiber links (solid lines). Right: The blow up shows that at the center of each connecting link, midway between any two nodes, there is an entanglement distribution center (blue circle with two blue dots) that fires entangled photon pairs out to two user nodes (grey circles with five red dots).[57]

57 Khatri, Sumeet, Corey T. Matyas, Aliza U. Siddiqui, and Jonathan P. Dowling. "Practical Figures of Merit and Thresholds for Entanglement Distribution in Quantum Networks." *Physical Review Research* 1, Article Number 023032 (2019) figure 1.a. https://journals.aps.org/prresearch/pdf/10.1103/PhysRevResearch.1.023032.

Figure 6.11, but rather the nodes are placed at random, as they are (more or less) in today's classical internet. The scale of this network is about the size of a few city blocks with the nodes spaced about 100 m apart. We do not need a quantum repeater over such a short distance. The entire quantum network itself is a distributed quantum repeater.

As seen in the figure, the network consists of user nodes that have quantum memories and entanglement sources that spew out entangled photon pairs at a very high rate – say a trillion ebits per second. We transduce the entangled photon pairs into two entangled qubits located in quantum memories at the user nodes. When this step is complete, the two neighboring user nodes share entangled qubits with each other in their memories. In quantum internet 3.0, the sources and user nodes are all small, low-powered, integrated optical devices. (We could even put the emitters on the user nodes if we wanted to, but this configuration in Figure 6.11 is a bit easier to analyze.) We see that the user nodes have multiple quantum memories (five each in this case) that share ebits with five other users on five nearby nodes. We also integrate entanglement swappers into the user nodes – adjacent to the memories. The swappers are also tiny chip-sized devices.

Let us now suppose that Alice wants to share an ebit with Bob. As before, in the star network, they are not directly connected, and so there is some switching to be done. Unlike the star network, where all switching is done at a central switching station, here, every user is also a switcher. This configuration is a great deal like the current classical internet, where a large router on an office building floor is in charge of switching data packets to the intended user on that floor. Also, like in the classical internet, Alice and Bob send out a request across the classical network (running in the background) that they would like to share entanglement. All the nodes between them report back with data usage rates (some nodes are busier than others at different times), and the network as a whole decides on the best route to entangle Bob's qubits with Alice's. The choice will depend on the time of day and whether or not some of the nodes are malfunctioning. The network chooses the route that gets Alice and Bob their entanglement as quickly and cheaply as possible. In Figure 6.12, I show two potential paths the system might select.

As in the classical network, computers and regular communications do all this while running in the background – Alice and Bob never see any of this – they just get a message that the entanglement swapping is complete. This procedure is like getting a note on your email client that

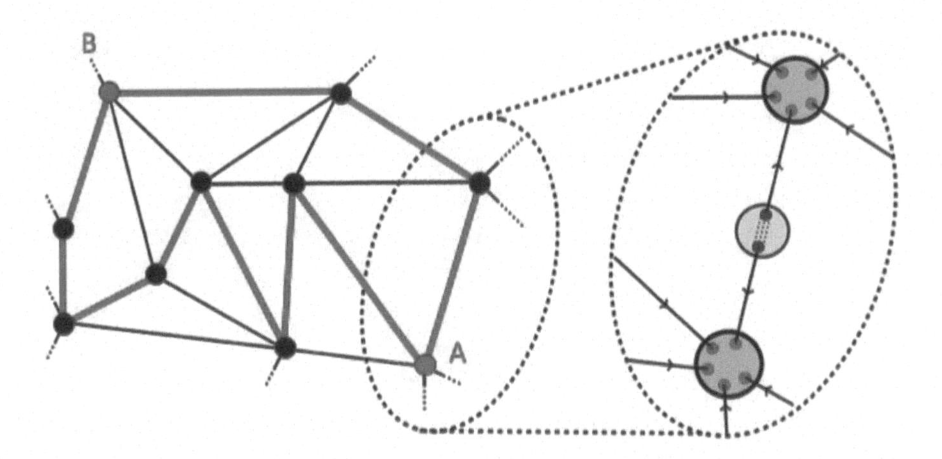

FIGURE 6.12 Alice wants to share an ebit with Bob, but Bob is not one of her nearest neighbors on the network. They will have to use the intermediate nodes as entanglement-swapping switches. I show two possible entanglement-sharing routes (out of tens of possible ways) with the thick red lines.

an email has arrived. Sometimes it comes instantly, and sometimes it takes minutes, all depending on the distance between the users and the route the packets take.

Let's make a segue here for a bit to describe the classical internet and how it works. Consider the network on the left of Figure 6.12 and ignore the blown-up quantum inset on the right for now. Imagine for now that this is a classical fiber network. The dots are the user nodes, and the links are the fibers. Alice writes an email and sends it to Bob. What happens? The email is first broken down into bits and bytes and assembled into small packets. I like to think of these parcels as tiny fish. The head of the fish contains the destination of the packet – Bob's laptop in this case. It also includes instructions on how to merge with all the other packets when it arrives in Bob's inbox. The tail of the fish contains a portion of the email itself.

After Alice's computer turns the email into packets, the system then makes many copies of these packets and sends them in a swarm across the internet. The figure shows two possible routes (out of about ten) the packets could take – the bottom zig-zag route and the more straightforward upper route. For redundancy, the many copies of the packets take all of the possible paths. The software disassembles the email into packets. It makes many copies of each packet, and all the packets – original and copies – are released into the network. This design gives the classical network *two* modes of robustness against nodes going down or links being severed, or nodes and links being overloaded with too many packets. The first level of robustness comes from the multiple routes and the second from the numerous copies of the packets.

The copies of the packets can take alternate routes – there are about ten in total. So even if half of the packets get lost or stalled on the upper route, due to network issues, their copies can keep going on the lower route. The robustness comes from having many routes and many copies of the packets. After some time, the packets arrive at Bob's laptop. Bob reassembles the email by first looking at the packet headers – the heads of the fish. From that information, he can discard extra copies of the packets and see if any packets are missing. If some are missing, he can send a short message back to Alice – asking her to resend them. Once Bob has all the packets, using the data in the packet headers, he can then reconstruct the entire email. Alice and Bob are unaware of any of this – their email client software is doing this in the background.

Let's now replace this classical network with an identical quantum network and see what is different. The quantum network can still utilize the multiple paths to increase the odds that the transmission will succeed, but it cannot make multiple copies of the quantum states due to the no-cloning theorem. The way to get around this copying problem is not to send the quantum data states themselves but rather send entangled pairs instead. Then, in analogy to the classical network, the users can send lots and lots of entangled pairs through the web – the pairs contain no information, but they are a resource like the electrical power in the old phone lines. The quantum network will distribute many ebits to Alice and Bob over multiple paths. Just like some classical packets get lost, some quantum ebits get lost. However, the network just keeps trying over and over again using different routes until it succeeds in entangling Alice and Bob. Once Alice and Bob accumulate enough ebits in their memory, then they begin quantum communication, such as teleportation, quantum key distribution, or distributed quantum computing.

In Figure 6.12, treating it now as a quantum network, I show two possible routes by which Alice and Bob can share an ebit by multiple entanglement swapping at intermediate nodes, that is, the intermediate nodes are the switches. Once the network decides what the best route is, it issues a command for all the intermediate nodes along the selected route to carry out entanglement swaps with their nearest neighbors along the way. Unlike the classical internet, where the data travels sequentially from node to node to node – on the quantum network – the intermediate switching nodes can carry out the swapping in any order that they like. They can all swap at once, or they can do it one-by-one down the line, or they can do it in random order – it doesn't matter. Recall though that the switches are always broadcasting classical bits with the results of their Bell measurements to the neighboring swappers. Those classical bits are required to complete the swapping. Since classical signals move at the speed of light, and the nodes are only 100 m apart, the system completes the entire chain of swapping in a few tens of microseconds. When finished, Alice and Bob now share many ebits – each composed of two entangled qubits, stored in their respective quantum memories. They can now use this for quantum crypto or teleportation or distributed quantum computing.

Going from the star network to the random network has a few critical advantages. In the star, all communications had to go through the central switch. If that switch went down or became overloaded, then all communications halted. In the random network, there is no central switch – every node is a small switch – and so there are many, many, many different routes on which we could implement the entanglement-swapping chain. In Figure 6.12, I show two possible chains. There is the simple upper chain that requires three links with two swaps. There is then the much more complicated zig-zag lower chain that requires six links and five swaps. You might think that the lower chain is much worse because the entanglement has to go much farther but remember the swaps exchange entanglement almost instantaneously – limited only by the arrival time of the classical signal bits. True, the longer you go, the more likely an error will creep in – but that is what the entanglement purification is for. If we run the full-blown quantum repeater protocol, fixing photon loss as well, the longer route does not require many more resources than the shorter route.

What the random network buys us is robustness against terrorist attacks, node failure, and overloaded switches. For the star network, there was only one path from Alice to Bob. For a sizeable random network, there can be billions. Even in the simple network in Figure 6.12, there are about ten different routes between Alice and Bob; if any node fails on any one circuit, you have another nine you can use. A random network is, provably, the most robust against loss of switching nodes. Even if you don't lose a node, one node on the upper route may be overloaded with traffic, and the network then optimizes the routes with respect to the internet traffic and chooses the route that is the least used, which is perhaps the zig-zag lower route. This sort of real-time choice of routes is what the classical internet does, but with the vital difference here that we are distributing entanglement and not data. What does that mean? Alice and Bob are constantly storing ebits in their memories, as before, until they have many gigaqubytes worth of them. At that point, the entire quantum network could go down, and they would still be able to use their ebits to carry on doing quantum communication as if nothing happened – so long as the classical backup network is still up and running to assist them.

The quantum internet 3.0 then looks quite a bit like the current classical internet. Fiber links connect users all over the earth via landlines. The satellites will all be phased out; we will still need a few for situations where you can't run fibers such as in ship-to-shore communications – the Chinese are experimenting with that now. We could even replace the satellites by entanglement-distributing drones for small areas, such as in a battlefield.

What about entangling two mobile phones? That is a bit of a tall order. We can't have everybody running around with long fibers attached to their phones. Once we make the quantum memories small, low-power, and cheap, everybody will have one on their phone. When you plug your phone into your charger, which you have to do every day or so, you get it recharged with electricity from the power grid. In the future, the charger will have a separate port to recharge your phone's quantum memory from the quantum-entanglement-distribution grid. When you unplug it, the phone now is charged and has a trillion qubits that are entangled with some other nearby node in the quantum network. Let's suppose Alice and Bob want to secure their text messages with QKD – the E91 protocol. First, their phones use their classical communication links to order the quantum network to allow them to share ebits. The network implements a series of swappers until it entangles all of Alice's qubits with all of Bob's. Once that step is completed, then Alice and Bob communicate directly over the classical phone link and execute the E91 protocol. They carry out measurements on their qubits, periodically interspersed with Bell tests, and communicate their measurement settings with each other. In a few milliseconds, Alice and Bob share a random key that is a trillion bits long. The quantum steps are now over, and then they implement a one-time pad to communicate securely with each other over the classical cell-phone towers.

When will this happen? I was discussing this idea with my colleague, Mark Wilde, a world expert in quantum information theory. He thought not for 100 years. I give it 10. Regardless, once this full-fledged quantum internet is in place, what on Earth shall we do with it? The U.S. Department of Defense developed the ARPANET to share data between large mainframe computers. Researchers invented the world-wide web to share scientific documents between labs. Who could have predicted then that, one day, the internet would be used for nearly all banking transactions, playing video games with strangers, and sharing millions of photos of a woman screaming at a cat? We had to first build the classical internet and start using it before we could figure out what it was suitable for. The same goes for the quantum internet.

In the following sections, I will lay out some ideas I have for using the full-blown quantum internet. I've mentioned some of these before – encryption, distributed quantum computing, and networks of quantum sensors. However, just like the classical internet, we really won't figure out what the quantum internet is useful for until it is up and running, and young people, like my high school students, begin playing around with it.

I HAVE A CODE IN MY NODE!

While I cannot predict everything that quantum internet 3.0 will be useful for, I can give you some ideas of what we think it will be helpful for. In this section, we focus on quantum random number generation and quantum key distribution. Both of these applications are very similar, and the requirements for implementing them on the network are nearly identical. I discussed both the quantum random number generator of the third kind and the Ekert 91 quantum key distribution scheme (E91) in some detail in the last chapter. That explanation only extended to one person trying to generate random numbers or two parties trying to share a secret code. Things change a bit on the net.

For both of these technologies, we require good sources of quantum entanglement and Bell measuring devices. Not required, but helpful, are good quantum memories. Quantum internet 3.0 has all of these. Let's start with E91, and I'll back out of that the random number generation. As we can see in Figure 6.12, after executing many steps, Alice and Bob share trillions of ebits in their respective quantum memories. They will now use these ebits to set up a shared one-time pad key, by converting their trillions of ebits into trillions of shared random zeros and ones.

Once they have these shared random numbers, they then can use them to encode and decode classical messages to each other. The cyphertexts cannot be hacked by any means – even by a quantum computer. What quantum-key distribution buys you is the ability to establish the secret key in such a way that nobody has a third copy of it. In quantum internet 2.0, the keys were transmitted from Alice to Bob using the BB84 protocol. There, uncertainty and unreality protect the transmission from eavesdroppers.

In E91, quantum entanglement – the third ingredient of quantum technology – is added to the mix. This extra feature gives us the means to prove, in a fundamentally different way, that nobody has a copy of the key. Alice no longer transmits the key to Bob, but rather the network magically generates the shared key on-sites from the ebits themselves. That is, in BB84, Alice creates a key and sends it to Bob. However, in E91, somewhat like teleportation, Alice and Bob carry out a procedure that generates the key instantaneously out of the ebits – pretty hard to tap that! One might image many ways to intercept a BB84 key that Alice is sending to Bob, but how do you snatch a key that does not yet exist? This feature makes E91 unique. If Alice does not *transmit* the key, but instead, Alice and Bob *create* the key out of *nowhere*, there is no way an eavesdropper can steal the key in transmission. The best the eavesdropper can do is spoof the entanglement distributors in the network. She could do this by replacing the entangled photon pairs with classically correlated photon pairs so that when Alice and Bob made key, the eavesdropper would know what it was. But several quantum and classical roadblocks are in Eve's way. To pull off the spoof, Eve would have to take control of *every* entanglement source in the network between Alice and Bob. That, Mack, would be some hack! Even if Eve could pull that off, Alice and Bob still can detect her. Recall that part of the E91 ritual is for Alice and Bob to sacrifice some of their ebits to continuously run a Bell test in the background. Recall the test spits out a number. If the number is two (or less), then there is an eavesdropper or trickster on the web. If the number is higher than two, then the network is secure.

In the BB84 trunk-line between Shanghai and Beijing, there is no such test for security – armed guards secure the rooms with the relay nodes at each train station. If the Chinese were to switch from BB84 to E91, replacing the relays with entanglement generators, then the guards would no longer be needed – the Ghost of Heisenberg guards the nodes.

(He cannot be bribed, and he does not draw a salary.) This entanglement advantage is called device-independent quantum-key distribution. The users can certify the security of their shared key using a Bell test. They don't need to have any idea of what is going on in the network. Nor do they need to know what makes up the sources and swappers.

To complete this section, we recall from the previous chapter that a certifiable quantum random number generator also uses shared ebits and Bell measurements to verify the randomness. Once Alice and Bob share random numbers, they don't *have* to use them to make a cryptographic key; they could instead sell them to a casino to secure its slot machines. Thus, the quantum network is not only a distributed quantum-key generator, but it is also a distributed random-number generator – there's an app for that too!

TELEPORTATION AND DISTRIBUTED QUANTUM COMPUTING

Twenty years ago, when I worked at NASA, I installed some software on my desktop called SETI at Home.[58] The Search for Extraterrestrial Intelligence (SETI) is a privately funded project that scans the skies with vast arrays of radio telescopes, looking for radio signals that could be coming from a planet with intelligent life on it. The telescopes take in way more data than can be processed on any one supercomputer, so the idea of SETI at Home is to turn all the laptops and desktops on Earth into a single gigantic supercomputer to analyze all that data. This setup is called networked computing – the network of computers forms a super-duper computer that is far more powerful than any of the individual machines. SETI at Home is using the unused processing power on millions of desktops and laptops – for free! All SETI at Home does is provide the software that divvies up the data processing into small jobs, which it then sends out across the internet, and then collects the processed data back again. While it has not found intelligent life yet, the thing is still running. There was a similar networked program that looked for enormous prime numbers and – unlike SETI at Home – that networked supercomputer *did* discover new prime numbers.[59]

These networked computer systems had their origin in the Beowulf Cluster, designed by my colleagues, Thomas Sterling and Donald

58 To my surprise, you can still do this https://setiathome.berkeley.edu/.

59 Finding prime numbers at home is more likely than finding aliens at home www.mersenne.org/.

Becker, at NASA. Their idea was that you could take cheap old PCs lying around in your building, IBMs, Macs, whatever, and hook them all together into a supercomputer using nothing but free software and a fast switch. The switch and the software would take the computer job and split it into little jobs spread over all the PCs with the more powerful machines getting more work to do and the least powerful less. The switch and software then take all the results from the component computers and compiles them into a single answer. The networked prime finder works the same way, except now the PCs are distributed around the world - not just around the lab. Of course, with a classical network, the central switch is sending back and forth packets of zeros and ones. To network quantum computers, you must instead send back and forth *quantum* states. How do you do that? You teleport them!

From 2011 through 2013, I was part of an IARPA-funded project "Protocols, Languages and Tools for Resource-Efficient Quantum Computation". The project was supposed to run for four years, but IARPA terminated it in May of 2013, giving us only 30 days advance notice.[60] There were several other teams besides ours, and we all had independent approaches for designing architectures, compilers, and software for several quantum computing platforms. In any computer design, one of the first questions you need to answer is, what is the most efficient way to move data around on the computer? On the classical computer, there is only one answer - you move the bits and bytes around on wires. On the quantum computer, there are two answers - you can either move the qubits or qubytes around on wires or you can teleport them. All of the research teams independently reached the same conclusion - the most efficient way to move quantum data is to teleport it. It does not matter if we are talking about a single quantum computer or a network of quantum computers, the answer is still teleportation. Why does teleportation win in the quantum case?

We have seen why in the discussion above. For a classical network, Alice can always make copies of the data, and so if the transmission line is faulty or has a lot of loss or noise, you can use the redundancy of the copies to overcome those problems. The no-cloning theorem prohibits such an approach in a quantum network. Since you can't make copies, the analog of the classical solution is to send the quantum states over

60 The historical trend would suggest that IARPA never met a quantum-computing project that it didn't like - *to kill!*

a quantum communication channel and hope for the best. A quantum communication channel, at the very least, allows you to transmit the quantum superposition state of an $|H\rangle + |V\rangle$ polarized photon, and not let that superposition collapse into a classical mixture of either H or V. That is, the quantum channel transmits qubits and the conventional channel only bits. Since we cannot copy the qubits, we have to send them in this superposition of an $|H\rangle + |V\rangle$ photon down the quantum channel, from transponder to transponder, until it gets to its final destination. Without the ability to copy it, it can only take one of the ten possible routes shown in Figure 6.12. If there happens to be too much traffic or a break on that route, you lose your state. If the state is carrying part of a computation from one quantum computer to another, the loss means the entire distributed quantum calculation fails, and you have to run the whole thing all over again. Luckily, as all the IARPA teams discovered, pre-shared entanglement, coupled with teleportation, fixes all that.

Not only is teleportation a more efficient use of resources over direct-state transfer, it comes with an additional feature. A bit can take on values zero or one (but not both). A qubit can be prepared in an arbitrary superposition of a zero and one. In principle, that superposition needs an infinite number of classical bits to describe it. If Alice has a *known* qubit and wants Bob to construct an identical qubit, in the worst-case scenario she would have to send him an infinite number of bits in order for him to duplicate her qubit. (This duplication does not violate no-cloning, which only applies to *unknown* qubits.) Instead, Alice can teleport the qubit to Bob, with all its infinite number of bits, at the cost of only one ebit (to initialize the teleportation) and two classical bits (to complete the teleportation). That is, by exploiting teleportation, an infinite number of bits is compressed into two bits and one ebit – how's *that* for data compression?

The quantum redundancy is not in having multiple copies of the transmitted quantum state, but instead having numerous copies of the pre-shared entanglement. If you think of entanglement as the electrical power on the landline, you would have to cut *all* the different circuits connecting you to the phone company, before you lost the ability to make a call. Entanglement is the resource that allows for quantum communication, just like AT&T's power grid is the resource that will enable you to make a call. Since entanglement is not the state that we are trying to send from Alice to Bob, but rather the resource that allows us to transmit the state, it is not so precious as the state itself. Again, in the landline analogy, AT&T can lose as much as 50% of the electrical power between its substation and your house (due to loss in the wires). But even with only 50% power, you can still get 100% of your phone calls through. *Compare this to the alternative scenario where AT&T drops half of all of your calls.* The power loss is not nearly as consequential as the phone-call loss itself. In the quantum analogy, the entanglement loss is not nearly as crucial as the transmitted-state loss. To repeat once more, the redundancy of having multiple copies of the classical data packets, on the classical internet, is replaced with the redundancy of having multiple copies of ebits on the quantum internet.

Using entanglement distribution, combined with entanglement distillation and purification, Alice and Bob can take a long time and take many tries to build up trillions of really high-fidelity ebits in their memories. Losing lots of ebits is not a problem. Losing the quantum states to be transmitted is the problem. Once Alice and Bob share lots and lots of ebits, then they can carry out distributed quantum computing over the network as follows. Let's suppose that Alice has software on her end that splits up the quantum-computing program into two parts – one part will run on her quantum computer and the other on Bob's – it's parallel computing in parallel universes! To get half of her quantum states over to Bob, she cannot just measure them and transmit the results to him classically. The measurement collapses all the states and destroys the quantum computation entirely. She must send him his share of the states quantumly. Using the ebits, one ebit per qubit state to be transmitted, she teleports all the states to Bob. The states arrive unaltered at Bob's location and the distributed quantum computation continues unabated. When he completes his part of the calculation, he teleports his output states back to Alice. Once Alice garners all these lovely quantum states together, she compiles Bob's output states with her states and only then makes the final measurement that spews out the prime factors of a trillion-digit number.

I can now design a new quantum-network problem solver. Instead of classically finding the biggest prime number, the distributed quantum computers will hack the largest public crypto key by factoring it into its composite primes. Of course, there is other stuff that we can do on a network of such quantum computers, such as quantum machine learning and quantum chemistry simulations. There are way more applications than that, but we won't know what they are until we build the quantum network, and my high school students start writing q-apps – quapps!? – to run on it. One of these anticipated quapps is a network of quantum sensors, which we shall discuss in the next chapter.

CLASSICAL COMMUNICATION OVER THE QUANTUM NETWORK

In all of the discussions above, I have assumed that we still have the classical internet running in the background. First, that is needed to send the classical bits to complete teleportation and swapping, and secondly, when entanglement is a precious resource, it is a waste of that resource to send classical data over a quantum network. But what happens when entanglement is a free resource? It turns out that you can transmit classical data over a quantum network at much higher speeds than you can over a classical network. Given this is the case, there will come the point where we will use the quantum network to transmit *everything* – from the states of a quantum computer to the recipe for my grandmother's Irish soda bread.

The quantum network is composed of quantum channels – the fiber or free-space links between the nodes. A classical channel can only transmit zeros and ones, but a quantum channel can transmit, also, superpositions of zeros and ones. Thence, you can use the quantum channel to send classical bits by just using the classical-to-quantum encoding $0 = |0\rangle$ and $1 = |1\rangle$. That's a dumb way to do it, as there is no quantum advantage – your data-transmission rate will be the same as if you had used a classical channel in the first place. With free entanglement, we can do better. Before the invention of teleportation, there was super-dense coding.[61] In Figure 6.13, we show the setup for super-dense coding. The idea is that if entanglement is freely available, then you can send two classical bits over a quantum channel with the help of one ebit. Buy one bit – and get the second one free! Super-dense coding is in many ways similar to teleportation. Indeed, teleportation was invented by some folks playing around with super-dense coding. As shown in the figure, the ebit source on the left of the figure produces the two-qubit ebit state $|\Phi^{+}\rangle_{AB} = |0_A\, |0\rangle_B + |1\rangle_A\, |1\rangle_B$ and sends half to Alice and half to Bob. They can either store the ebit in their memories or perform the protocol on the fly.

61 Super-dense coding https://en.wikipedia.org/wiki/Superdense_coding.

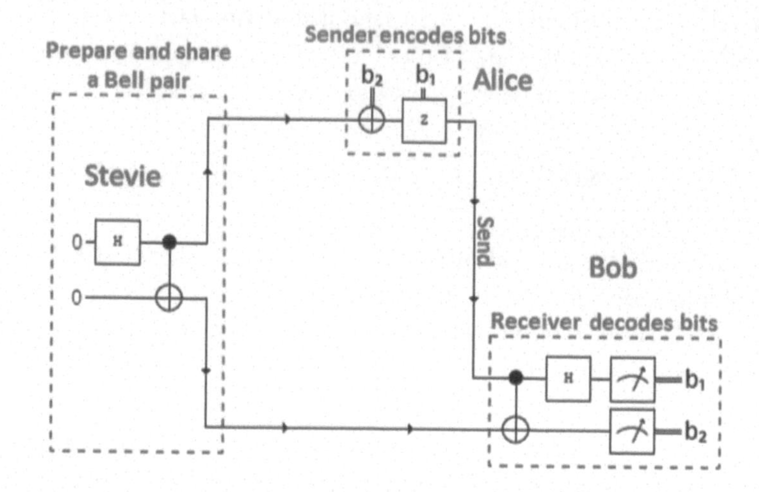

FIGURE 6.13 Super-dense coding. Stevie (left) sets up a shared ebit between Alice and Bob. Alice has two classical bits a_1 and a_2 that she wants to send to Bob. Alice encodes her classical bits, b_1 and b_2 quantumly and then uses two quantum computer gates to entangle her bits with her half of the ebit. Bob then takes his half of the ebit and performs two additional quantum-computing gate operations. When he is done, out pops Alice's two bits in his lab.[62]

Alice (A) has two classical bits, a_1 and a_2, that she would like to send to Bob (B). There are four possible messages she can send with two bits, 00, 01, 10, or 11, which correspond to a_1 equals 0 or 1, and a_2 equals 0 or 1. That's just the amount of classical information needed to complete a teleportation protocol! Alice then uses the naive classical-to-quantum encoding $0 = |0\rangle$ and $1 = |1\rangle$.

Alice then uses two quantum gates to entangle her two quantum-encoded bits with her half of the ebit. She then sends her half of the ebit, thus modified, to Bob. Bob then performs two more quantum gates and – hey presto! – out pops Alice's two classical bits. If the ebits are free and the gates are free, then they transmit two classical bits for the price of one, thus doubling the classical transmission rate. Here's where it gets weird. Bob can't extract the classical bits unless he has both of qubits in the ebit. An eavesdropper cannot tap this classical communication channel unless she has access to both Alice and Bob's labs. In

62 Wilde, Mark M. *Quantum Information Theory.* Cambridge University Press, Cambridge, UK, figure 7.3. Used with permission from the author who is the copyright holder.

addition to doubling the transmission rate, super-dense coding is a type of quantum cryptosystem. Alice's two bits are the plaintext, and the encoding into the ebit is the cyphertext.

To illustrate this point more clearly, imagine the ebit source is in Bob's secure lab. He sends half of the ebit to Alice and keeps the other half in his memory. Alice performs the encoding and sends the modified half of her ebit back to Bob. Bob then retrieves her two bits of information. The only way an eavesdropper could access those two classical bits is if she took control of Bob's lab. Hence, super-dense coding can be used as a cryptosystem to either directly send encrypted messages or to send one-time-pad keys for later use. We get double the transmission rate and quantum security for the price of an ebit. Can we do better? The answer is that we can do *way* better.

In quantum information theory, there are many approaches to send classical information over a quantum network.[63] In the most general case, the end-users, Alice and Bob, share entanglement, the classical bits are encoded into entangled quantum states at Alice's end, and the receiver at Bob's end also exploits entanglement. There are no simple formulas for how well you can do with this general scheme, but if entanglement is free, then you can send *lots* more classical data over a quantum network at a much higher rate than you can by using a classical network. The super-dense coding factor of two is the lower limit on how well you can do. The upper limit depends a great deal on who is entangled with what. The take-away message is that if you have a free quantum network, and you want to send classical data at a very high rate, you use the quantum network to do it. The classical network will eventually fall to the wayside, like copper wires in the old telephone landlines, and eventually everything will be transmitted quantumly. We may still want to keep the classical internet around as a backup system – in the same way that I cling to my copper-wired telephone landline – to use in case of the Apocalypse.[64]

63 Wilde, Mark M. *Quantum Information Theory*. Cambridge University Press, chapter 20. https://arxiv.org/abs/1106.1445.

64 In that same 2018 conference at U.S. Global Strike Command, I asked the Docs, "Do you have some sort of backup communication system to use once World War III is over?" The Docs whispered to each other and then cryptically replied, "Well, Prof, we have a *thing*." I recoiled a bit and retorted, "A thing! What kind of thing?" More whispering and then came this chilling response, "Well, Prof, it depends on *what* survives."

CHAPTER 7

Networks of Quantum Sensors

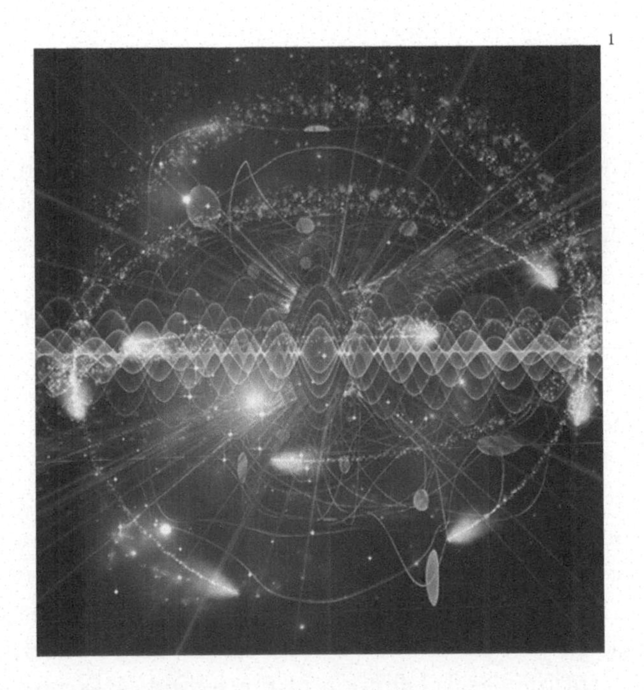[1]

'A miracle has happened, and a sign has occurred here on earth, right on our farm, and we have no ordinary pig.' 'Well,' said Mrs. Zuckerman, 'it seems to me you're a little off. It seems to me we have no ordinary spider .' 'Oh, no,' said Zuckerman. 'It's the pig that's unusual. It says so, right there in the middle of the web.'[2]

1 Image produced by U.S. Government employees in the line of their work and is not subject to copyright https://casis.llnl.gov/seminars/quantum_information.

2 White, E.B. *Charlotte's Web*, 85–87. Harper Collins, 1952. http://www.worldcat.org/oclc/881367088.

QUANTUM SENSORS

In addition to using the quantum internet for security and distributed quantum computing, the artificial intelligentsia are paying a great deal of lip-service to quantum networks of quantum sensors. To understand what that means, first, I'll describe what a quantum sensor is, and then tell you what we gain from networking them. As I mentioned above, a quantum sensor and a quantum computer are very similar things. The computer uses quantum effects to solve intractable math problems and the sensor to measure things more accurately than is possible classically. I have written several review articles on quantum sensors such as the very readable paper "Quantum optical metrology – the lowdown on high-N00N states,"[3] and the more technical paper "Quantum optical technologies for metrology, sensing, and imaging."[4] To be a quantum sensor, the device must utilize at least the first two requirements for a quantum technology – randomness and unreality. For most modern quantum sensors, we throw entanglement in for good measure. I'll describe here in some detail a quantum atomic clock, which measures the passage of time, and then I'll argue that all other quantum sensors are similar.

As the name would suggest, an atomic clock is composed of atoms, and the atoms are the qubits. We'll continue our notation using atomic spins to represent the qubits. The manipulation of the atoms that we use to make the clock are identical to single-gate operations that we perform in a quantum computer. Hence, quantum atomic clocks also make good quantum atomic computers. We show the most general representation of a qubit in Figure 7.1.

Using the notation $|0\rangle = |\uparrow\rangle$ and $|1\rangle = |\downarrow\rangle$, we map the abstract qubit in Figure 7.1 to the physical spin of the atom qubit that makes up the atomic clock. For concreteness, we use a cesium atom, which is used to give the current universal definition of the second. A second is defined to be *exactly* 9,192,631,770 periods of rotation of the qubit around the equator in Figure 7.1. Now we have to figure out how to put our qubit onto the equator in the first place. We do this with something called a Ramsey interferometer, which I sketch in Figure 7.2.[5]

3 Dowling, Jonathan P. "Quantum Optical Metrology – The Lowdown on High-N00N States." *Contemporary Physics* 49 (2008): 125–143. https://doi.org/10.1080/00107510802091298, https://arxiv.org/pdf/0904.0163.pdf.

4 Dowling, Jonathan P., and Kaushik P. Seshadreesan. "Quantum Optical Technologies for Metrology, Sensing, and Imaging." *Journal of Lightwave Technology* 33 (2015): 2359–2370. https://www.osapublishing.org/jlt/abstract.cfm?uri=jlt-33-12-2359, https://arxiv.org/abs/1412.7578.

5 Ramsey Interferometer https://en.wikipedia.org/wiki/Ramsey_interferometry.

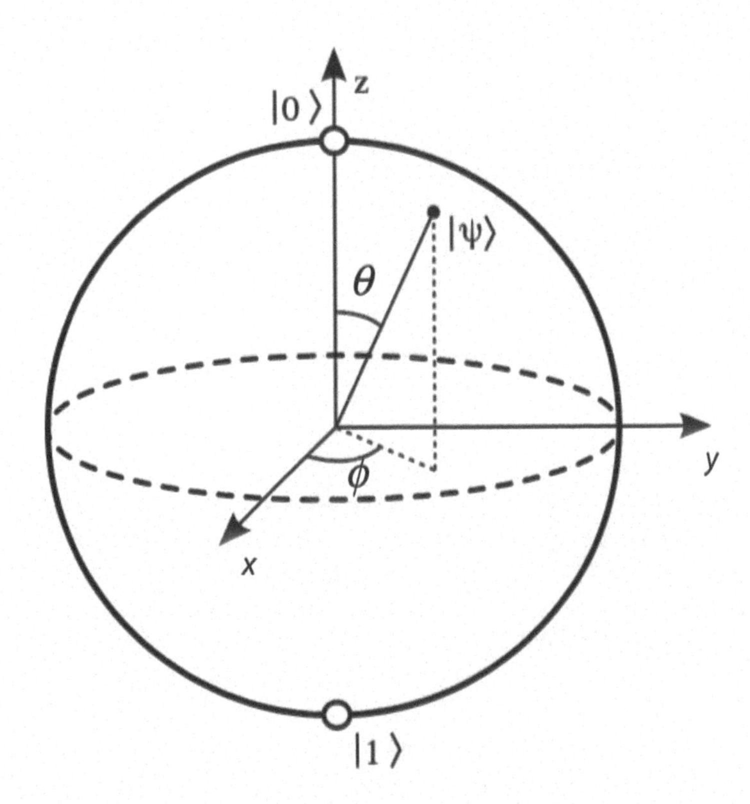

FIGURE 7.1 A representation of the qubit on a sphere. The classical bits 0 and 1 are shown on the north and south poles, respectively. However, the qubit state, which is an arbitrary superposition of 0 and 1, is represented by the arrow $|\psi\rangle$. The latitude angle θ and the longitude angle ϕ completely specify the qubit's state.

As we see in Figure 7.2, the cesium-atom qubits emerge from a source on the left in the spin-down state, which is the south pole of the sphere in Figure 7.1. The first microwave pulse kicks the qubit onto the equator, where it begins to rotate at a rate of 9,192,631,770 revolutions per second. After some time (the time that we wish to measure), a second microwave pulse is applied that collapses the qubit either into spin up or down. The ups are detected by the upper Pacman device, and the downs by the lower device. Think of the arrow rotating around the equator as the hand of a clock. The arrow makes 9,192,631,770 complete rotations per second. That means if you time one rotation, you can measure time down to one part in 9,192,631,770 – which works out to about ten billionths of a second. That's why atomic clocks are so accurate. It is easy to use the

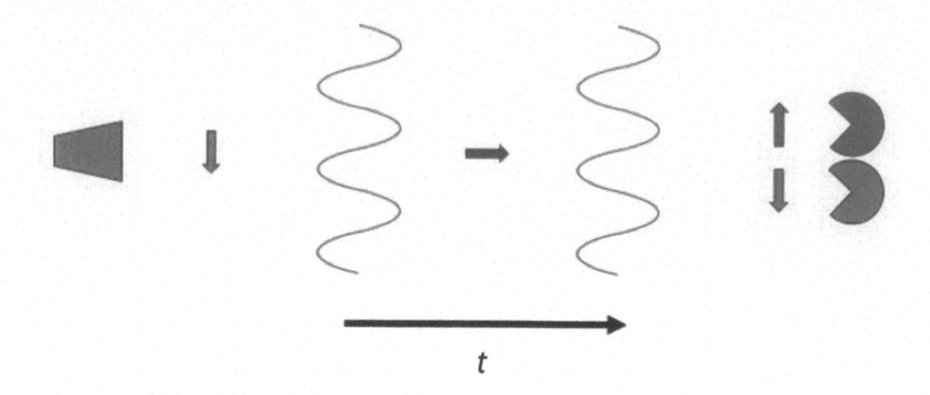

FIGURE 7.2 A schematic of a Ramsey interferometer for measuring time. The clock-qubit atom in the state $|1\rangle = |\downarrow\rangle$, is emitted from an atom source on the left. The first wiggly line is a microwave pulse that puts the qubit in the state $|0\rangle + |1\rangle = |\uparrow\rangle + |\downarrow\rangle = |\rightarrow\rangle$ on the equator in Figure 7.1. The qubit then begins to rotate around the equator at the rate for a cesium atom. After a preassigned time, we apply a second pulse that collapses the qubit into either $|\uparrow\rangle$ or $|\downarrow\rangle$.

interferometer to measure one rotation or fractions of a rotation. The probability that the spin hits the upper detector, minus the probability that it hits the lower sensor, is a function of the time to be measured. This function oscillates up and down as time evolves, as shown in Figure 7.3.

These oscillations are called Ramsey fringes and are like the numbers on a clock. Each oscillation corresponds to one rotation of the qubit around the equator, which itself corresponds to measuring 10 ns. With this contraption, we can accurately measure the passage of time. This clock is a quantum technology that uses quantum unreality (the superposition of up *and* down qubits between the microwave pulses) and quantum uncertainty (the collapse of the qubit after the second microwave pulse). The thing that we are measuring is that longitudinal angle ϕ, which is directly proportional to the time elapsed. The better we can estimate that angle, the better we can measure the passage of time. This longitudinal angle ϕ is called the phase of the sensor – a term that we'll use again and again. But, for now, think of it simply as the angle of rotation of the qubit on the equator.

In order to get good signal to noise, the time is allowed to vary, as shown in Figure 7.3, and we experiment with many qubits. The measurement of

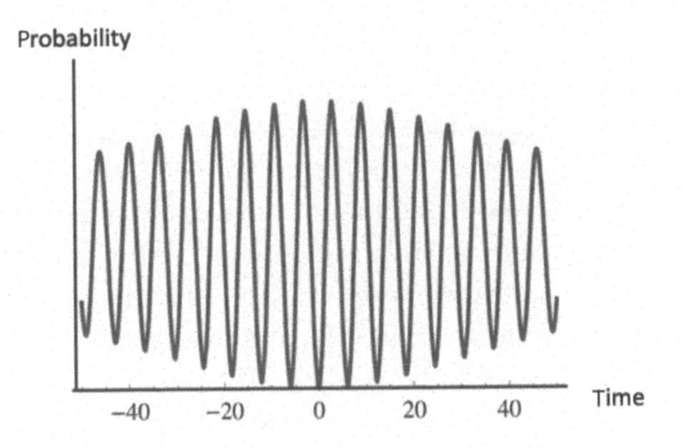

FIGURE 7.3 The probability difference between the upper and lower detectors in Figure 7.3. The peak-to-peak width of each oscillation corresponds to one rotation of the qubit arrow around the equator in Figure 7.2.[6]

each qubit gives a data point on the oscillating curve, and it is the entire curve that we are trying to estimate. The more qubits we use, the more data points we have, and the better our estimate of the curve. Think of timing a three-minute egg with a clock. If you look at the clock every few *minutes*, then you will have either an overcooked or an undercooked egg. If you look at the clock every few *seconds*, then the egg will be cooked nearly perfectly. The more times you look – the more data you take – the better your estimate of the passing time. However, the accuracy or uncertainty in the time goes down slowly with the increasing number of qubits. If we use four qubits, and the accuracy (in units of nanoseconds) is ±1/2, then if we use 100 qubits, the accuracy improves only to ±1/10 – not to 1/100 – as you might expect. We call this behavior the law of large numbers, and I discussed it at length in my previous book.[7] The law tells is that taking more data is a good idea, but also that you have to take an agonizingly lot of data to get even a small improvement. Here's where entanglement comes to the rescue.

6 Pollock, J.W., V.I. Yudin, M. Shuker, M.Yu. Basalaev, A.V. Taichenachev, X. Liu, J. Kitching, and E. A. Donley. "AC Stark Shifts of Dark Resonances Probed with Ramsey Spectroscopy." *Physical Review A* 98, Article No. 053424 (2018), Fig 2. This work is a contribution of NIST, an agency of the U.S. government, and is not subject to copyright https://doi.org/10.1103/PhysRevA.98.053424.

7 Law of large numbers https://en.wikipedia.org/wiki/Law_of_large_numbers.

We replace the first microwave pulse with a sequence of pulses that implement a quantum-computing circuit. The circuit takes the first four atoms and puts them into an entangled state, $|\uparrow\uparrow\uparrow\uparrow\rangle + |\downarrow\downarrow\downarrow\downarrow\rangle$. The four atoms are in a superposition of all four pointing up and all four pointing down. (This is an example of the GHZ state discussed previously.) The four arrows combine to make one big arrow that swings around the equator four times as fast as before. After some time passes, we apply a second series of pulses that implements a quantum circuit that reverses what the first circuit did. This stratagem collapses the four-atom state into either $|\uparrow\uparrow\uparrow\uparrow\rangle$ *or* $|\downarrow\downarrow\downarrow\downarrow\rangle$. If the uncertainty of the measurement with the four individual qubits was ±½, then the uncertainty with the four entangled qubits is ±¼. It follows that if we do the same with 100 entangled qubits that the uncertainty goes from ±1/10 to ±1/100. Entanglement gives us a quadratic improvement in the signal to noise.

We can map all quantum sensors to this atomic-clock scheme. In all cases, there is a qubit that rotates around the equator by an angle ϕ. In the case of the atomic clock, the angle is proportional to time, and so the interferometer measures the passage of time. For different quantum sensors, the angle contains information about other things – such as a magnetic or electric field, a gyroscopic rotation rate, acceleration, a gravitational field, temperature – or the signal from a passing gravitational wave. All quantum sensors have this same form. If the math is the same – then the physics is the same! Note that when we introduced entanglement into the sensor, we immediately had to use the language and tools of quantum computing to explain and implement it. Again, we see here the close connection between quantum sensors and quantum computers – a quantum sensor is a special-purpose quantum computer. That's why quantum sensors fall under the topic of quantum information science in the U.S. National Quantum Initiative.

The qubit is measuring the angle ϕ that carries the signal we are trying to sense. That angle is encoded into the quantum state of the qubit. In the above sensor scenario, we allow the qubit to rotate – to sense – and then we make a measurement that collapses the qubit into a data point. What if we were trying to measure the time or gravitational field over the entire country of Japan? We could just measure the results for each sensor at each location in Japan and then average them – the Japanese do this with seismometers to sense impending earthquakes. This approach works fine, but we are then still stuck with the law of large numbers, where here, the numbers are the number of sensors (and not the number of qubits in the sensor). Once again, classically averaging the outputs of 100 sensors gives an uncertainty of only ±1/10 and not ±1/100. What if we could coherently combine or even *entangle* the disparate sensors together? To do that, we would have to take the qubits in each sensor and teleport them to a central data-processing center, or we would have to entangle all the sensors together in the first place. If we have the quantum network 3.0 already in place, then we can hook all the different sensors up that way, and then quantumly combine all of their qubits to get a vast improvement in signal to noise. That's what we will talk about next.

COHERENT SENSOR ARRAYS

Near Socorro, New Mexico, there is a Very Large Array of giant radio antennas called, somewhat unimaginatively, the Very Large Array. (See Figure 7.4.)

FIGURE 7.4 The Very Large Array of radio telescopes. Each telescope is about the size of a house.[8]

8 Image from Shutterstock under a license from Taylor & Francis https://www.shutterstock.com/image-photo/radio-telescope-view-night-milky-way-1110702323.

When imaging distant astronomical objects with a telescope, there are two important metrics of goodness for the telescope – the amount of light it gathers and the resolution of the distant image. You want the image to be bright and all the small features in it easy to see. You can improve both of these features of the telescope by making the diameter of the receiving dish bigger. It is obvious that a bigger telescope gathers more light, but it is less obvious that it also improves the image quality. This latter property is called the angular resolution of the telescope.[9] The rule that – the bigger the telescope, the better – also applies to cameras. You don't see the professional wedding photographer taking photos with his mobile phone – instead, he lugs around a huge camera with a big lens on the front. That big lens pulls in more light and improves the resolution of the photo. Each telescope in the Very Large Array is about the size of a house. To improve your image quality, you could try to make a single telescope the size of a football field. However, the current radio telescopes, which make up the array, are almost at the point where they would collapse under their weight. The largest, single, radio telescope in the world is the great Arecibo Radio Observatory, which has a total area of about 20 football fields. The builders solved the weight problem by nestling the giant telescope into a huge sinkhole crater in Puerto Rico.[10] That thing is a wonder of engineering and the end of the line for making telescopes bigger and bigger. Fortunately, there is an end-run around this rule. Instead of making a single large telescope – you phase-lock together a whole bunch of smaller telescopes into one single virtual telescope whose effective diameter is equal to the diameter of the array. The Socorro array has a layout shown in Figure 7.5.

Each telescope in the array has a diameter of 25 m, but the width of the entire collection is about 21 km. The idea is that we build one, giant, virtual telescope 21 km across, out of all the smaller telescopes. The virtual telescope has the same diameter as the entire array, and thus this thing is much larger than the Arecibo Observatory. To construct this virtual telescope, you can't just take the images from each telescope and average them – that takes you back to the law of large numbers and a crappy photo. You must treat each telescope as a quantum sensor that measures the phase of the light coming from the astronomical object. To get the big telescope, you must coherently add all the phases from

9 Angular resolution https://en.wikipedia.org/wiki/Angular_resolution.
10 Arecibo Observatory https://en.wikipedia.org/wiki/Arecibo_Observatory.

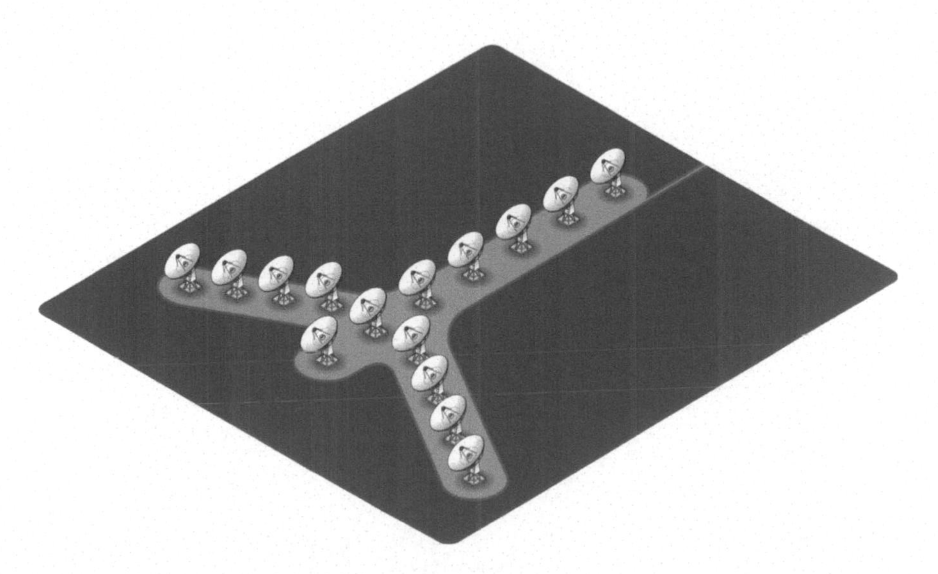

FIGURE 7.5 A schematic of the layout of the Very Large Array. The array consists of 28 telescopes – each with a diameter of 25 m. The size of the array is about 21 km across.

the small ones – that is, the virtual telescope is a giant interferometer. As we have seen, an interferometer is a quantum computer, and that's how we bring in quantum networks. I show a schematic of a two-telescope interferometer in Figure 7.6.

If you squint really hard – can you find the qubit in Figure 7.6? Well, it is there. Consider a single photon arriving from the star. The photon moves like a classical wave, and its wave function spreads out much bigger than the distance between the telescopes. The photon is in a superposition of taking the longer path to A and the shorter route to B. Hence, at the point where the photon passes the astrometric baseline, we can write its state as $|0\rangle_A|1\rangle_B + |1\rangle_A|0\rangle_B$. That is, one photon going to B with nothing going to A *and* one photon going to A and nothing to B. We can then rewrite this in qubit notation using $|0\rangle_A|1\rangle_B = |\downarrow\rangle$ and $|1\rangle_A|0\rangle_B = |\uparrow\rangle$, which gives us the qubit state $|\downarrow\rangle + |\uparrow\rangle = |\rightarrow\rangle$. This state is precisely that in Figure 7.2. Because the path to Alice is longer than the way to Bob, the qubit rotates around the equator at a longitudinal angle ϕ, which is proportional to the path difference. The beam splitter/combiner is equivalent to the second microwave pulse in Figure 7.2, and the adjustable delay is equal to the tunable time

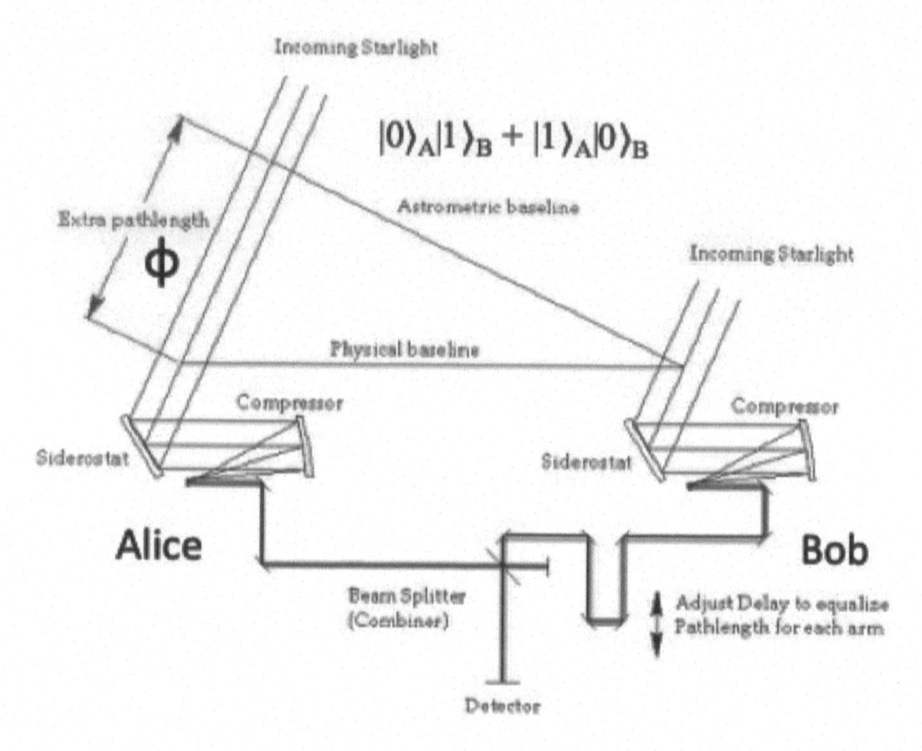

FIGURE 7.6 A schematic of a two-telescope virtual array. Light from a star comes in from the upper right. Alice and Bob have two spatially separated telescopes. The light impinging on the left-most telescope has to travel a bit farther – the extra path length. This path-length difference means that the signals to the telescopes have a phase difference ϕ.

between pulses. The first pulse is generated by the photon propagation being split into two paths – the first beam splitter is the sky. Even though it does not look like it, the device is a quantum interferometer like the atomic clock. Here, the angle ϕ is measured, instead of the time, and using a Fourier transform; you can back out the image of the star from repeated measurements of the angle.[11]

For imaging at radio wavelengths, the telescopes do not have to be in direct contact. Radio waves have a very slow period of oscillation – so slow that you can record their intensity patterns at each telescope, and then copy it to a disk with a timestamp for each data frame. If you have

11 Astronomical optical interferometry https://en.wikipedia.org/wiki/Astronomical_optical_interferometry.

an excellent atomic clock, then the timestamp is sufficiently accurate to allow you to interferometrically combine all the signals from all the telescopes later, at some central station, and extract the high resolution at your leisure. However, this approach becomes problematic for short-wavelength astronomy in the infrared or visible part of the spectrum. Light of these wavelengths oscillates a million times faster than light at

FIGURE 7.7 "ALMA image of the protoplanetary disc around HL Tauri (which is 450 lightyears from the earth). This is the sharpest image ever taken by ALMA – sharper than is routinely achieved in visible light with the NASA/ESA Hubble Space Telescope. It shows the protoplanetary disc surrounding the young star HL Tauri. These new ALMA observations reveal substructures within the disc that have never been seen before and even show the possible positions of planets forming in the dark patches within the system." Photo credit ALMA (ESO/NAOJ/NRAO).[12]

12 Image from https://commons.wikimedia.org/wiki/File:HL_Tau_protoplanetary_disk.jpg. This image was produced by the ALMA Observatory. The images and videos distributed from the public ALMA websites (www.almaobservatory.org, www.alma.cl, and kids.alma.cl) along with the texts of press releases, announcements, pictures of the week and captions, are licensed under a Creative Commons Attribution 4.0 International License, and may on a non-exclusive basis be reproduced without fee provided the credit is clear and visible. See https://www.eso.org/public/outreach/copyright/.

radio wavelengths, and the atomic-clock method does not work. The telescopes have to be hooked together directly into a single optical interferometer, such as the one shown in Figure 7.6. Just as with the Very Large Array, you can do this with more than one telescope. The record of carrying out such interferometry with superb synchronized clocks is held by the Atacama Large Millimeter/submillimeter Array (ALMA). It is located in Chili on a chilly wind-swept plain – high in the mountains.[13] As the name would suggest, it images astronomical objects at wavelengths of millimeters – far shorter than the Very Large Array radio waves – but still considerably longer than a telescope operating with visible light.

The problem with physically phase-locking two optical telescopes is that it is extremely challenging to do over long distances. All the beam paths shown in Figure 7.6 have to be aligned, and using the synchronized-clock trick won't work, because the atomic clocks are not precise enough for the high-frequency oscillations of visible light waves. The photo of the protoplanet formation in Figure 7.7 was taken using wavelengths of millimeters. The dark rings are made by newly forming planets plowing through the dust cloud surrounding the central star. The image is in false color, since millimeter light is far below infrared light, and so it is not visible to the naked eye. Visible light has wavelengths on the order of hundreds of nanometers, which is a million times shorter than millimeter waves. If we could do this same type of telescope interferometry in the optical domain, then we would have nearly a million times better resolution in the image of Figure 7.7. We would be able to see the planets themselves and not just the rings that they make. We would even be able to carry out spectroscopy on those planets' atmospheres to look for oxygen – a tell-tail signal for the existence of life. But without atomic clocks that are million times better than what they use with ALMA, what are we to do? There is a great future in teleportation. Think about it. Will you think about it?

The notion is that, if we cannot connect the optical telescopes physically, and if we don't have those really good clocks, we could instead teleport the phase angle ϕ from one telescope to the other over the quantum network.[14] If there are many telescopes, then each telescope

13 ALMA Observatory https://www.almaobservatory.org/en/home/.

14 Gottesman, Daniel, Thomas Jennewein, and Sarah Croke. "Longer-Baseline Telescopes Using Quantum Repeaters." *Physical Review Letters* 109, Article No. 070503 (2012). https://doi.org/10.1103/PhysRevLett.109.070503; https://arxiv.org/abs/1107.2939.

measures its angle ϕ and we teleport all these different angles to a central processing station. There we interferometrically combine them in a special-purpose optical quantum computer. All we need to do is hook all these telescopes up to the quantum internet and then teleporting those phase angles to a central quantum computer is a cinch. Imagine thousands of telescopes distributed all across the surface of the earth – all phase-locked together via the quantum network.

The resolution of the smallest features in the image – planets or death stars – goes like λ/D, where λ is the wavelength of the light, and D is the diameter of the telescope array. To resolve things as small as possible, we want to make this ratio as little as possible. Hence by moving from millimeters to nanometers in wavelength, we improve the resolution by a million. The ALMA array has 66 mm-wave antennas spread out over a distance D = 16 km. Suppose, instead, we have a vast network of optical telescopes operating in the hundreds of nanometers regime. Further that the network has the diameter of that of the entire Earth, namely, D = 13,000 km. We get another factor of 1,000 improvements in the resolution from this diameter. With such a giant telescope – we could see features a billion times smaller than ALMA can. Imagine if we improved the resolution of the photo in Figure 7.7 by a billion! Let us suppose that one of the dark rings in the picture is caused by an earth-sized planet, which has a diameter of around 10,000 km. Since we can see the dark circle, ALMA can resolve things about 10,000 km across. Now we switch to the Giant Earth Quantum Optical Observatory (GEQuOO). Take 10,000 km, divide it by a billion, and you get – hold on to your hat! – one centimeter! With this earth-sized quantum-telescope array – I could see alien slugs slithering across the surface of that far-distant planet – 450 lightyears away!

Although here in this section, I have discussed only regular telescopes, you can replace those with any astronomical observatory that measures a phase. The two famous blackhole finders, the Laser Interferometer Gravitational-Wave Observatories (LIGOs), are giant optical interferometers that have a configuration somewhat like that in Figure 7.6.[15] They also measure a phase angle, but this time that angle corresponds to the signal of a passing gravitational wave. The first detection of such a wave occurred in 2015 and won the Nobel prize in physics for three people in 2016. Since January of 2019, the LIGOs have become quantum sensors.

15 LIGO https://en.wikipedia.org/wiki/LIGO.

Just as we discussed with the clocks, the LIGO-istas have managed to entangle together some of the laser photons that circulate in the interferometer. This quantum trick allows them to beat the classical limit in sensitivity by a factor of two. That may not seem like much, but it means that they can see twice as many gravitational wave sources – typically inspiraling black holes – in the north-south direction, twice as many in the east-west direction, and twice as many in the up-down direction. The final result is that they can see 2 × 2 × 2 or eight times as many wave sources as before. We have one of the LIGOs just about a 40-minute drive from my university (Figure 7.8).

Each LIGO is much more like an antenna than it is like a telescope. When you get a signal on your car radio, the radio wave from the station

FIGURE 7.8 The LIGO interferometer in Livingston, Louisiana. The two long arms at right angles to each other contain two laser beams that go out and bounce back, interfering at the central command center. (The thing not at right angles is a service access road.) Each arm is 4 km long. Photo courtesy of the LIGO collaboration.[16]

16 Images are provided by the LIGO Laboratory. LIGO Lab images may be used freely in the public domain with appropriate acknowledgment https://www.ligo.org/multimedia/gallery/llo.php.

strikes your car antenna and causes the electrons in the metal to oscillate up and down. The radio then converts these oscillations into sound. Similarly, when two distant black holes are locked into a death spiral, they emit gravitational waves – when these waves hit the LIGO antenna, they cause the length of the arms to shrink and grow in an almost periodic fashion.[17] The interferometer measures this oscillation and converts it into a signature of the gravitational wave, a simulation of which is shown in Figure 7.9. In this figure, we see the oscillations of the LIGO arms as a function of time. Initially, the signal looks periodic, but then as the two holes begin to merge, the oscillations start to speed up. With that speed up, the wavelength consequently becomes shorter (Figure 7.9).

It is a general principle of antenna theory that the best receiver should have an antenna length as close to the wavelength of the signal as possible. Thus, in Figure 7.9, just before the holes collide (far right) the wavelength is the shortest – around 1 km. Therefore, the LIGOs, with their 4-km-long arms, are perfectly designed to see these inspiraling compact astrophysical objects. However, the LIGOs are not good at seeing gravitational waves with much longer wavelengths. The LIGO collaboration can use the two

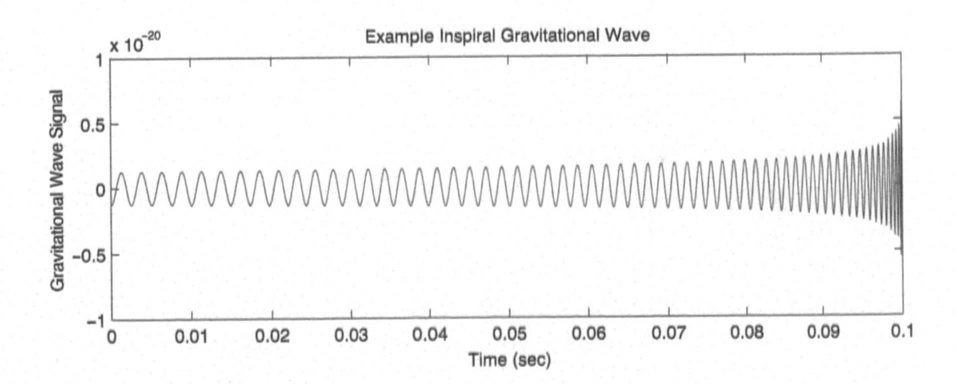

FIGURE 7.9 Simulation of the waveform of an inspiraling binary blackhole. As the two black holes begin to merge (far right), the frequency of oscillations speeds up, and the wavelength gets shorter. (Credit: A. Stuver, LIGO Lab/Caltech/MIT.)[18]

17 Here is a cool video of a simulation of two black holes in a death spiral https://en.wikipedia.org/wiki/Binary_black_hole.

18 Images are provided by the LIGO Laboratory. LIGO Lab images may be used freely in the public domain with an acknowledgement to LIGO Lab/Caltech/MIT https://www.ligo.org/multimedia.php#images.

LIGOs to triangulate the location of the source. Suppose, in analogy – you have two cars with radio antennas and excellent atomic clocks on board. By timing the difference between when identical radio waveforms arrive, the two vehicles could triangulate the location of the radio station. LIGO does this same thing to pinpoint the location of the source in the sky. But this process is all done classically – the two LIGOs are not phase-locked like our two telescopes in the previous example. Here to the rescue, once again, is the quantum internet.

The idea is similar to that of the telescopes. Each LIGO antenna measures a phase angle ϕ, which changes in time and carries information about the gravitational wave. The two LIGO antennas, the other one is in Washington State, are about 4,000 km apart. Using the quantum internet, we teleport the two phases to a central processing quantum computer, as before, and combine them interferometrically. When we are done, we have a single LIGO with a 4,000 km arm. To get the second arm, we need to bring the European detector VIRGO on board. This machine would allow us to see signals with a wavelength of 4,000 km, which theory predicts correspond to gravitational waves produced in that infernal maelstrom of the big bang.[19]

CLOCKS IN SPACE

Folks ask me all the time, "How do I ever use Einstein's theory of relativity, or Schrödinger's quantum theory in my real life?" Well, nobody has ever asked me that – but I wish they would. When that day comes, I have an answer, "You use them all the time with your smartphones!" In a world of omnipresent mobile phones, everybody has access to the Global Positioning System (GPS). Each time you quiz Google Maps or Apple Maps or Waze for directions to the closest gas station or Thai restaurant – you summon the ghosts of Einstein and Schrödinger! In your phone, there is a little antenna that is constantly scanning the skies for radio signals from GPS satellites. Once it locks onto at least two of those spacecraft, then it uses the information in those signals to triangulate your location on the map. With two satellites, you get your latitude and longitude; with three, you also get

19 Romano, Joseph D., and Neil. J. Cornish. "Detection Methods for Stochastic Gravitational-Wave Backgrounds: A Unified Treatment." *Living Reviews of Relativity* 20 (2017): 2. https://link.springer.com/article/10.1007/s41114-017-0004-1.

altitude, and with four, you get nanosecond-precision timing. Many scientists and engineers used the GPS signals for that accurate timing information. If the receivers are not moving around, say strapped to your roof, then they can average the timing signal and get down to even picosecond precision. Buying a GPS monitor is much cheaper than buying and maintaining your own atomic clock. What do Albert and Erwin have to do with it?

Let's start with Herr Einstein. Each GPS satellite has a very accurate atomic clock on board – just like the atomic clock we discussed above. A few times a day, each spacecraft synchronizes its orbiting clock to a master clock on the ground. In this way, we synchronize all the flying clocks with each other. However, due to the effects of relativity, the clocks in orbit run at different speeds than their identical cousins on the ground. To keep the whole thing working, we have to upload computations that compensate for those Einsteinian effects. If we did not do that, then your Waze app would be off by hundreds of meters – and it would tell you to steer your car into a ditch. The Ghost of Einstein is your Guardian Angel. The Ghost of Schrödinger is in charge of all the clocks. Each clock is a Ramsey interferometer, and therefore a quantum technology.

Each satellite broadcasts a radio signal that is unique to that specific satellite. The radio signal contains two pieces of information – where the satellite is and what the time is. That is the information that your smartphone extracts. Once your phone gets the signals from two different satellites, that is enough to triangulate your position. Add more sats and, to boot, you get altitude and the precise time.[20]

---

20 Fifteen years ago, when I was shopping for a house here in soggy old Louisiana, I carried with me a folder of flood maps and my Garmin GPS receiver. I did not look at any houses listed in a flood zone on the maps, and I refused to buy one that was less than 40 feet above sea level. (Many houses in Baton Rouge are at or below sea level – protected only by the levee system.) Upon arriving, with my real estate agent, at a house to inspect, I would first place the Garmin on the front steps of the house and then tour the place. This pause gave the Garmin time to average the signal and give a precise measurement. Remember, classically averaging data is a slow process. Upon exiting the house, I would pick up my Garmin and look at the elevation. More often than not, it was under 40 feet, and I would report to my agent, "This house is unacceptable." My behavior drove her nuts. "Why don't you tell me this *before* we tour the home? We could skip the house and save a lot of time." I would then explain to her that I needed to integrate the satellite signal, and for good measure, I would give her a lecture on the law of large numbers. After I finally bought a house with her – she quit her job! But I was right. In 2016, Baton Rouge experienced a terrible flood, and all the neighborhoods around mine flooded. My neighborhood was almost completely spared – and it turned out that my house had the highest elevation in the neighborhood.

The GPS would work even better if the earth had no atmosphere – gasp! The turbulence in the air messes up the radio signals, which limits the accuracy of a receiver on a moving platform to about ±3 m. That's good enough to get you off at the right exit on the freeway, but if your plane does not know the altitude of the runway any better than that, you're in for a rough landing. The issue is the clock synchronization – the timing signals in the radio waves are being buffeted about by the wind, and this effect limits the timing accuracy to a few nanoseconds. What if you could teleport the timing information past the atmosphere? This idea is something our group proposed two decades ago.[21] Now, with the rise of quantum networks, scientists are taking another look at the idea. Recall the atomic-clock qubits. We measure the time by measuring that phase angle ϕ. If Alice and Bob want to synchronize two clocks, they need to lock the phase on one clock to that on the other. The two qubits need to be whirling around the equator at the same speed and in synchrony. When one qubit hits the prime meridian, the other qubit must hit that same meridian at the same time. The current method for synchronizing the GPS clocks to the ground clocks is equivalent to us directly transmitting the state of Alice's clock qubit to Bob's. When you send the qubit through the atmospheric turbulence, it messes up the phase-angle rotation rate. But we have a fix for this! Let's make Bob on the satellite part of the quantum network. (See Figure 7.10.)

Bob's spacecraft has a source of entangled photon pairs. He sends one to Alice and converts one into a qubit in his memory. Alice then takes the other photon and converts it to a qubit in her memory. The ebit is *also* messed up by the atmosphere, but we can share a lot of them and then deploy entanglement purification to get a few purer ebits. This setup is just like the quantum internet 3.0. Don't directly send your precious quantum state – build up exquisite shared entanglement instead – and then teleport the state. We may use the same process to synchronize all the clocks on the ground via the quantum internet. In the most far-out scenario, the clocks in space will share entanglement with each other, and we won't need so many space-to-ground links. In principle, we could synchronize one space clock with one ground station, and then use the

21 Jozsa, Richard, Daniel S. Abrams, Jonathan P. Dowling, and Colin P. Williams. "Quantum Clock Synchronization Based on Shared Prior Entanglement." *Physical Review Letters* 85 (2000): 2010. https://doi.org/10.1103/PhysRevLett.85.2010; https://arxiv.org/abs/quant-ph/0004105.

FIGURE 7.10 Artist's concept of a world clock consisting of multiple spaceborne atomic clocks interlinked via quantum entanglement. Credit: The Ye group and Steve Burrows, JILA.[22]

space-to-space links to synch up all the other space clocks with that ground-linked one.[23]

DISTRIBUTED QUANTUM SENSORS

In the above, we discussed how to improve the sensitivity of a sensor by entangling the qubits. What we have not discussed is how to improve the sensitivity of a network of sensors. You do that by entangling the sensors together over the quantum internet. Let's start by modeling the quantum sensor as a quantum-computer gate array, as shown in Figure 7.11.

The figure is an abstraction of the atomic clock in Figure 7.2. Each qubit is independent of the others – no entanglement – and the sensing precision is ±1/2. Now we have the source entangle all the qubits together, as shown in Figure 7.12.

22 Image from https://jila.colorado.edu/yelabs/news/sky-clocks-and-world-tomorrow.

23 "Sky Clocks and The World of Tomorrow." https://jila.colorado.edu/yelabs/news/sky-clocks-and-world-tomorrow.

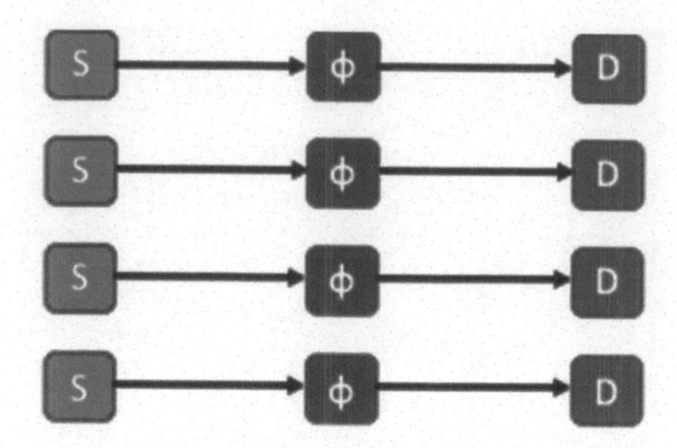

FIGURE 7.11 Quantum computer model of an unentangled quantum sensor. The four sources (S) each emit one qubit in the state $|0\rangle + |1\rangle = |\uparrow\rangle + |\downarrow\rangle = |\rightarrow\rangle$. Each qubit enters the sensing region ϕ and whirls about the equator by a phase angle ϕ, which is what we are measuring. In the detectors (D), an estimate of the phase angle is made from a measurement.

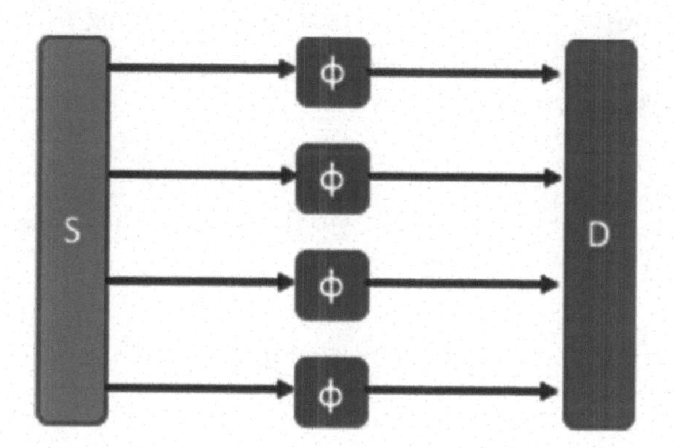

FIGURE 7.12 The source is now a quantum circuit that entangles the four sensor qubits together into the GHZ state $|\uparrow\uparrow\uparrow\uparrow\rangle + |\downarrow\downarrow\downarrow\downarrow\rangle$.

The source now is an entangling quantum computer circuit that prepares the four sensor qubits into the entangled GHZ state $|\uparrow\uparrow\uparrow\uparrow\rangle + |\downarrow\downarrow\downarrow\downarrow\rangle$. As we have argued above, we can depict this state as a single qubit whose state arrow is four-times longer than that of any individual qubit., $|\uparrow\uparrow\uparrow\uparrow\rangle + |\downarrow\downarrow\downarrow\downarrow\rangle = |\rightarrow\rightarrow\rightarrow\rightarrow\rangle$.

This new large qubit we can draw on the equator of its sphere, and the rotation rate changes from ϕ to 4ϕ. That is, the big qubit spins around the equator four times as fast as a single qubit does. That change in speed leads to a quadratic improvement in signal to noise, and so we can now measure the phase ϕ with a minimum uncertainty of $\pm 1/4$ instead of the $\pm 1/2$ that we got with the four unentangled qubits.

We imagine that we now have lots of these entanglement-enhanced sensors hooked up to the quantum internet. The state of the big qubit can be teleported just as easily as the state of the little qubit, and so we can pool the results and add all the different phases at some central station, as shown in Figure 7.13.

Each sensor sits at a node that shares ebits with the central station C. The phase information from each sensor is encoded into a qubit at each node and teleported back to C. Then, at C, we add the phases coherently. This is the scheme we used for the telescopes. However,

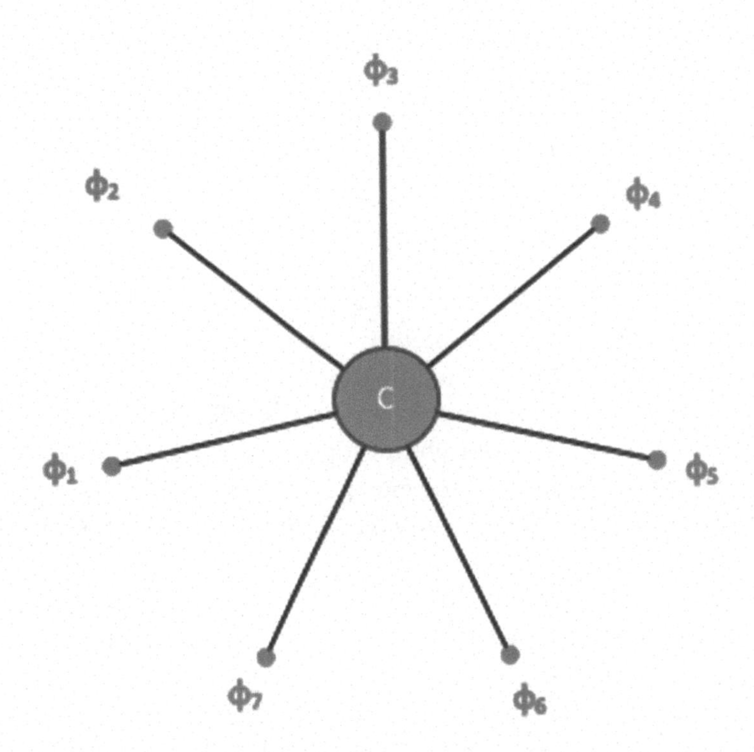

FIGURE 7.13 A star network of quantum sensors. Each sensor at the outer nodes measures its own signal, perhaps by using internal entanglement to get a better signal to noise. We encode the phases into qubits whose state we teleport to the central processing station C, where we combine them coherently.

since the individual sensors are not entangled with each other, the signal to noise improves only quadratically with increasing numbers of sensors. That is, if the precision which we obtain with 4 sensors is ±1/2, then the accuracy that we get with 100 sensors is only ±1/10. You get minimal improvement going from 4 to 100 sensors, because of that pesky law of large numbers. Nevertheless, such a sensor array could be quite handy, just like its sister, the telescope array.

In the eye of your mind, consider that the sensor array in Figure 7.13 is located in Greater Los Angeles. Los Angeles is very prone to earthquakes. Let's make a place at each node for an interferometric, fiber-optic strain sensor.[24] You put down the fiber on the ground where you want to measure the strain – the motion of the ground due to an earthquake. We send light down the fiber, and then it bounces back off a reflector, and then it comes back up the fiber. On its way back out, the light interferes with the light going in and produces an interference pattern that contains information about the total length of the fiber – a lot like LIGO. When earthquake waves pass by, the fiber stretches and shrinks in tune to the wave. The interference pattern moves around, and you get a time-dependent phase ϕ that carries information about the strength and frequency of the earthquake wave. If we classically average the signals from the seven different sensors in Figure 7.13, we can triangulate the location of the epicenter and estimate the strength of the earthquake.[25] This information is collected at the speed of light. The earthquake waves move much slower – at the speed of sound (in dirt). Hence, we can exploit the information to make an early warning system. The sensor array can call everybody's cell phone and tell them to take cover. If you are not right on top of the epicenter, this warning can arrive tens of seconds before you feel the quake. That's enough time to grab your cat and – a bottle of wine! – and fling yourself under your dining room table.

But what if we coherently link all the fiber sensors into a giant array? As with the telescope, we will get vastly improved resolution of the earthquake waves. We can then produce a real-time earthquake map that shows the direction and intensity of the waves in great detail. Given enough sensors, the early warning system could call your mobile phone and, using the GPS location finder on your device, tell you exactly when

24 Fiber optic sensor https://en.wikipedia.org/wiki/Fiber-optic_sensor.

25 "Dark Fiber Lays Groundwork for Long-Distance Earthquake Detection and Groundwater Mapping." *Science Daily* (2019), https://www.sciencedaily.com/releases/2019/02/190205151006.htm.

the earthquake will hit you and how bad it will be. With 20 s of warning, you have enough time to take the next freeway offramp before the highway collapses into a heap of rubble.[26] That being said, we are coming to the end of the road as it pertains to networked quantum sensors.

The last network I will consider is one where the sensors are entangled with each other and not just with the central station. Consider Figure 7.14.

In this figure, we allow the nodes to share a seven-qubit entangled state. Which seven-qubit entangled state? Nobody knows yet. Nevertheless, the goal is that, by exploiting this entanglement, we can measure each of these seven phases more accurately than by just averaging the results. We can beat the law of large numbers. We do this by using

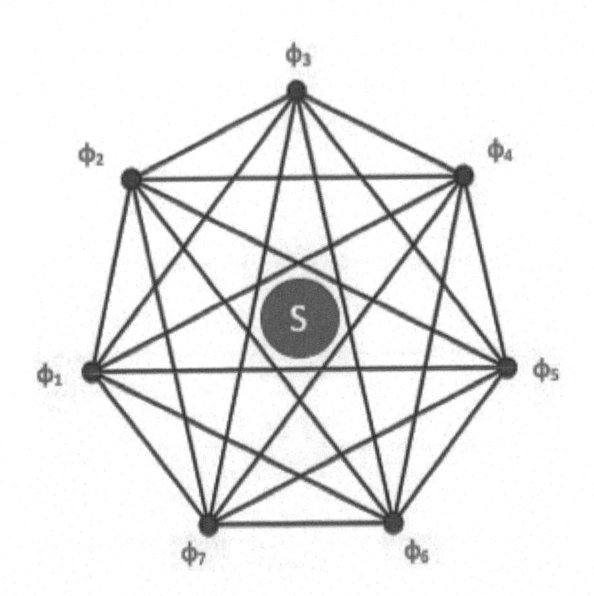

FIGURE 7.14 A completely connected star graph of sensor nodes. We hide the quantum internet, and we highlight the entanglement of each sensor with all other sensors. Stevie (S) at the source teleports out to the nodes a seven-qubit entangled state. The qubits in that state sense the seven phases ϕ. The entangled qubits, thus modified, are teleported back to Stevie for further processing.[27]

26 Taylor, Alan. "The Northridge Earthquake: 20 Years Ago Today." *The Atlantic* (17 January 2014); https://www.theatlantic.com/photo/2014/01/the-northridge-earthquake-20-years-ago-today/100664/.

27 Figure taken from https://commons.wikimedia.org/wiki/File:Complete_graph_K7.svg. It is in the public domain and not subject to copyright.

something called multiparameter estimation, where the parameters are the difference phases. In Figure 7.14, Stevie at the source teleports a seven-qubit entangled state out to the seven sensor nodes. Each qubit senses the phase ϕ of its respective node. Then the seven qubits, thus modified by the phases, are teleported back to Stevie. He then carries out a multiqubit detection along with some data processing. A schematic of the procedure (with just four nodes) we show in Figure 7.15.

Here we have only four qubits, for simplicity, and we separate the source (S) and detector (D) to avoid confusion. Work by our group and others has shown that if you send out the right four-qubit entangled state, then you can beat the law of large numbers and the quadratic scaling law.[28] If all four of the phases are the same, we know that the four-qubit entangled GHZ state $|\uparrow\uparrow\uparrow\uparrow\rangle + |\downarrow\downarrow\downarrow\downarrow\rangle$ is optimal. However, if the phases are different, then a multiparameter estimation needs to be used. Discovering which state is optimal depends on the number of qubits, and the state can be challenging to find.

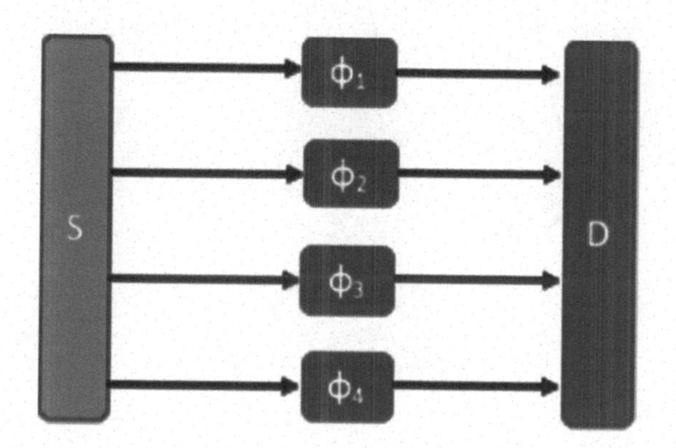

FIGURE 7.15 The source S sends out a four-qubit entangled state. Each qubit senses one of the four phases and then is sent to the detector D. The detector makes a joint measurement on all the qubits and extracts an estimate of each of the four phases.

28 You, Chenglong, Sushovit Adhikari, Yuxi Chi, Margarite L. LaBorde, Corey T. Matyas, Chenyu Zhang, Zuen Su, Tim Byrnes, Chaoyang Lu, Jonathan P. Dowling, Jonathan P. Olson. "Multiparameter Estimation with Single Photons." *Journal of Optics* 19 (2017); https://doi.org/10.1088/2040-8986/aa9133, https://arxiv.org/abs/1706.05492. I'm proud to report that Margarite LaBorde and Corey Matyas were my *undergraduate* students!

For example, for four qubits, there are lots of possible entangled states, which consist of adding together all the following possible basis states,

$$|\uparrow\uparrow\uparrow\uparrow\rangle, |\downarrow\uparrow\uparrow\uparrow\rangle, |\uparrow\downarrow\uparrow\uparrow\rangle, |\uparrow\uparrow\downarrow\uparrow\rangle, |\uparrow\uparrow\uparrow\downarrow\rangle, \ldots, |\uparrow\downarrow\downarrow\downarrow\rangle, |\uparrow\downarrow\downarrow\downarrow\rangle.$$

There are 16 such basis states in total. Combining the two on each end gives you back the GHZ state, $|\uparrow\uparrow\uparrow\uparrow\rangle + |\downarrow\downarrow\downarrow\downarrow\rangle$. Another relevant four-particle state, the W state, we construct via,

$$|\downarrow\uparrow\uparrow\uparrow\rangle + |\uparrow\downarrow\uparrow\uparrow\rangle + |\uparrow\uparrow\downarrow\uparrow\rangle + |\uparrow\uparrow\uparrow\downarrow\rangle.$$

It turns out neither of these is optimal for the four-parameter estimation. What we have been doing is using a computer optimization algorithm to try all possible four-qubit entangled states and then see which gives the best answer. The problem is that there is an infinite number of these, even with four qubits. You have to try all possible combinations of adding the 16 basis states together – and the states don't have to be equally weighted. Even with just four qubits, it takes our supercomputer days to find the optimal state. For seven qubits, as in Figure 7.14, there are 128 basis states. At that point, the optimization problem becomes intractable – the sensor network has become a full-blown quantum computer. You can't efficiently simulate a quantum computer on a classical computer! To find the optimal state for a large number of qubits will have to be done on a quantum computer – perhaps found with some tricky math that nobody has discovered yet. Regardless, it does seem that, under the right conditions, a multiparameter estimation can do better than a single-parameter estimation.[29] This result then implies we can get even better signal to noise on our earthquake warning system by entangling the sensors to each other. Distributed quantum sensing is a very new and active field of study.

29 There are lots of references to multiparameter phase estimation in this paper, Proctor, Timothy J., Paul A. Knott, and Jacob A. Dunningham. "Multiparameter Estimation in Networked Quantum Sensors." *Physical Review Letters* 120, Article No. 080501 (2010). https://doi.org/10.1103/PhysRevLett.120.080501, https://arxiv.org/abs/1707.06252.

Epilogue

[1]

They want to deliver vast amounts of information over the Internet. And again, the Internet is not something that you just dump something on. It's not a big truck. It's a series of tubes. And if you don't understand, those tubes can be filled and if they are filled, when you put your message in, it gets in line and it's going to be delayed by anyone that puts into that tube enormous amounts of material, enormous amounts of material.

United States Senator Ted Stevens (R-Alaska)[2]

1 www.shutterstock.com/image-illustration/email-tube-post-20055826.

2 Italics mine. See https://en.wikipedia.org/wiki/Series_of_tubes.

END GAME

Today, Friday, 27 December 2019, I'm sitting in my home office in my pajamas as I write this. I have been sitting here in my pajamas writing every day since classes ended three weeks ago. I only leave the house to pick up the mail, the newspaper, or to walk the dogs. Why? The deadline to have this book into the publisher is Wednesday, 2 January 2019 – less than a week. I am currently suffering from Quantum Writer's Zeno Paradox. As I approach the deadline for finishing my book, the number of people messaging me, "Have you finished the book yet!?" increases at an exponential rate – thus ensuring that I will never finish the book.[3] But this is the end. I now have to go back and proofread everything once again and then upload it to my server where the publisher can find it.

I had the idea for this book shortly after the publication of my last tome in 2013. Initially, I chose the title, *Schrödinger's Rainbow: Renaissance in Optical Quantum Interferometry*, but that seemed too technical. As with the last book, I started writing the semester I have off from teaching – a sabbatical I get every seven years. The *Oxford English Dictionary* defines "sabbatical year" to mean "… the seventh year, prescribed by the Mosaic law to be observed as a 'Sabbath' in which the land was to remain untilled …"

My first sabbatical was in the spring of 2011, when I started the first book, and my second was in 2018, when began writing this one in earnest. I wrote the proposal for this book in 2017. By that time, quantum networks had begun taking off, as Rodney Van Meter's book, *Quantum Networking*, had already been published in 2014. After each sabbatical, when I'm sufficiently untilled, I move into a new research direction. We held the annual 2017 Southwestern Quantum Information and Technology workshop in the spring here at my university, and there I began talking to other researchers in earnest about quantum networks. Motivated by these discussions, I started to write the book in May of 2017 – just as the Chinese began to publicly release the first earth-shattering Mozi satellite data.

By the time I wrote this book proposal, in the fall of 2017, I was all in and changed the name to the current title, *Schrödinger's Web: Race to Build the Quantum Internet*. The publisher sent my book proposal out to some of my colleagues to review, and one response was that it is too soon to think about the quantum internet. My timing, however, was perfect.

3 Quantum Zeno Effect https://en.wikipedia.org/wiki/Quantum_Zeno_effect.

This book should appear in the summer of 2020, right when the buzz about the quantum internet will be approaching the peak of its hype cycle.

I wrote this book to be a popular book – not meant for physicists. That means I have to explain complicated abstract concepts using analogies and as little math as possible. Inevitably, the parallels are not perfect, and in creeps a bit of ambiguity. I hope the general audience won't notice or care and will enjoy the book. If you did not read the footnotes, you should go back and read them now – that's where all the funny jokes and stories are!

As I did in my previous book, I spent a great deal of time reviewing the history of the science before diving into the application. The quantum internet exploits quantum states of light. What is light? I answered that in chapter one. What is quantum light? I partially explained that in chapter two. However, a critical and counterintuitive type of quantum light is an entangled pair of photons – I spent all of chapter three on that. Why am I doing this? Well, when you see the word photon or entangled state in the later chapters, I want you to have some physical intuition of what I am talking about.

In my field, I often hear the phrase, "Nobody understands quantum mechanics." I think this is attributed to Richard Feynman. I wouldn't say I like this statement. What it means is that quantum mechanics is counterintuitive – cats being simultaneously dead and alive and all that. But what is intuition? For me, intuition is the ability to grasp new ideas based on my past experiences. Whenever I drop a pencil, it always falls down and never up. Why? Well, that's just the way it is. If somebody drops a plate, my intuition tells me it will fall down. You can't have a hunch about physical processes, the like of which you have never encountered before. However, if you are used to working with photons in a superposition of horizontal and vertical polarization, then you're not going to be too surprised to find cats in a superposition of alive and dead. It was my goal, in the first three chapters, to build up your intuition on the nature of light and quantum states. I hope that when you got to chapters four and beyond, that intuition helped you wade the rapids of gushing quantum verbiage.

When designing a computer or network architecture, first, you define the elementary building blocks, like gates and such. I did this for the quantum photonic blocks in chapter four. In chapter five, I showed you how to put these together into more significant network elements. Chapter six was supposed to be the last chapter, on the applications of the quantum network, but the stuff on quantum sensors got too long, and so I split it off into a final chapter seven.

Having finished the book, you now know as much about the quantum internet (and its applications) as I do. If you have a hankering to learn all of this at a more technical level, I encourage you to get Van Meter's book – although it is a bit out of date and also a bit pricey. If you wait a year or less, some colleagues and I are working on a more up-to-date book, *The Quantum Internet – A New Frontier*. This comprehensive book is to be over 500 pages long, and the authors have started calling it "the doorstop".

Index

N

O

P